GCSE
Physics
QuickCheck Study Guides

John Avison
BSc, CPhys, MInstP
Head of Physics and Electronics,
Marton Sixth Form College, Middlesbrough

Jane Cartledge
BSc, MPhil, CPhys, MInstP, AFIMA
Deputy Head of Physics and Electronics,
Marton Sixth Form College, Middlesbrough

Nelson

Thomas Nelson and Sons Ltd
Nelson House   Mayfield Road
Walton-on-Thames   Surrey
KT12 5PL   UK

51 York Place
Edinburgh
EH1 3JD   UK

Thomas Nelson (Hong Kong) Ltd
Toppan Building 10/F
22A Westlands Road
Quarry Bay   Hong Kong

Distributed in Australia by

Thomas Nelson Australia
480 La Trobe Street
Melbourne   Victoria 3000
and in Sydney, Brisbane, Adelaide and Perth

© John Avison and Jane Cartledge, 1988

First published by Thomas Nelson and Sons Ltd 1988

ISBN 0-17-448151-9

NPN 987654321

Printed in Great Britain by Butler & Tanner Ltd,
Frome and London

Acknowledgements

We are grateful to the following examining bodies for
permission to reproduce questions from sample GCSE
papers: the London and East Anglian Group (LEAG);
the Southern Examining Group (SEG); the Welsh Joint
Education Committee (WJEC); the Northern Examining
Association (NEA), which is made up of the Associated
Lancashire Schools Examining Board, Joint Matriculation
Board, North Regional Examinations Board, North West
Regional Examinations Board, Yorkshire and Humber-
side Regional Examinations Board; the Midland Examin-
ing Group (MEG); the Northern Ireland Schools
Examination Council (NISEC); and the Scottish
Examinations Board (SEB).

Examination Boards

London and East Anglian Group (LEAG)
c/o University of London Schools Examination Board
Stewart House
London WC1B 5DN

Southern Examining Group (SEG)
c/o University of Oxford Delegacy of Local Examinations
Ewert Place
Summertown
Oxford OX2 7BZ

Welsh Joint Education Committee (WJEC)
245 Western Avenue
Cardiff CF5 2YX

Northern Examining Association (NEA)
c/o Joint Matriculation Board
Manchester M15 6EU

Midland Examining Group (MEG)
c/o Oxford and Cambridge Schools Examination Board
Elsfield Way
Oxford OX2 8EP

Northern Ireland Schools Examination Council (NISEC)
Beechill House
42 Beechill Road
Belfast BT8 4RS

Scottish Examinations Board (SEB)
Ironmills Road
Dalkeith
Midlothian EH22 1BR

Contents

Introduction

How to use this book

This book has been written to help you prepare for your GCSE Physics Examination. It will be of most use to you towards the end of your physics course and when you are ready to start revising for the exam. You will also find it useful when preparing for tests and exams during your GCSE course.

The book provides all the key ideas needed for your exam, gives you advice on how to study the topics, help in revising and preparing for the exam and also includes specimen questions to give you practice in answering exam questions. In addition, a list of the topics required for the different GCSE syllabuses is included in the form of a QuickCheck Syllabus Grid.

Using the QuickCheck Syllabus Grid

This book covers all the GCSE physics syllabuses, so there will be a number of topics in the book which will not be needed for your particular exam. You can check which topics are in your GCSE syllabus by using the QuickCheck Syllabus Grid. Select the topics you need by referring to the symbols in the grid:

- ● the whole topic is included
- ○ part of the topic is included

Where neither of these symbols appears against a topic, it means that the topic is not included in your syllabus.

The addresses of the examination boards are given before this Introduction. You can write to them for copies of past examination papers and the full syllabus for your course.

Studying the topics

- ● Read the advice on the following pages on how to prepare for an exam.
- ● Select your topics using the QuickCheck Syllabus Grid.
- ● Study the topics one at a time. Try to understand as well as to remember and work through the examples of calculations.
- ● Do the QuickCheck Review Questions which appear after each group of topics.
- ● Check the answers at the back of the book and go back to the topic to sort out anything you could not do or remember.
- ● Do the GCSE examination-style questions on pages 124 to 141.
- ● Check you answers to these questions against the hints and answer summaries given at the end of the book.

Review questions

QuickCheck review questions are provided after each small group of topics. These questions are closely linked to the work in each topic and allow you to test your knowledge and understanding topic by topic. Answer the review questions when you have studied each topic.

So that you can use the questions again for final revision, it is better not to write the answers in the book. Instead, write them out separately, along with any questions or diagrams that need completing.

When you have done all the review questions on a group of topics, look up the answers on pages 142 to 151. Go back to the topic notes to check on anything which you could not do, did not understand or forgot.

GCSE examination-style questions

On pages 124 to 141 GCSE examination-style questions are provided for all the topics in this book. Some of these questions are taken from specimen GCSE examination papers provided by the examination boards. Answering these questions will help you prepare for your GCSE exam in the following ways.

They will give you:

- ● a good idea of the kind of questions you will meet in your exam;
- ● practice in exam technique;
- ● a further chance to test your knowledge and understanding of physics;
- ● more practice at using the other skills needed for GCSE physics such as problem solving, applying your knowledge to new situations, using graphs and diagrams and handling data.

The answers to these questions are given on pages 152 to 157 and include:

- ● hints for, or summaries of, written answers;
- ● numerical answers, sometimes with hints to help you get the correct answer.

How to prepare and revise for the GCSE examination

The GCSE examination aims to test much more than just what you can remember. You must also develop certain skills, be able to apply your knowledge and be able to solve problems. Above all you must show that you **understand** your physics.

What skills and knowledge do I need for a GCSE physics exam?

- ● You must know and understand the important laws and concepts, definitions and formulas, units and symbols. Even in those exams where a list of formulas is given, you still need to know what the symbols stand for, and unless you understand them you will not be able to apply the formulas.

- ● You must know some key facts, e.g. the names, charges and relative masses of the particles in the atom. Such facts are often summarised in lists or in tables, e.g. page 115. You must learn these.

- You must understand how to use flow charts like those used to explain how energy changes its form, e.g. page 110. You must be able to work out your own energy flow chart for an unfamiliar example.

- You will need to be able to understand and draw diagrams. To explain an idea or illustrate an experiment, you should be able to draw a simple diagram and label all the important parts. You need to remember important diagrams, e.g. the diagram of the eye on page 45.

- You will need to be able to draw graphs and present data in tables and other forms of charts. See page 31 for examples of bar charts. You should also be able to explain what a graph shows and to obtain data from it. On page 93 the use of graphs of stretched materials shows what you should be able to do.

- Problem solving is a skill which you need in both practical physics and for the written exam. For the examination you must know how to solve numerical problems. This involves understanding how to use formulas and how to find the right information from the question. In this book you are shown how to use each formula by a worked example. For example, on page 22 you are shown how to use the equations of motion.

- Your practical skills will be tested by your teacher during the course. You need to be able to use equipment safely and correctly, to read scales, design experiments and solve practical problems.

How can I learn?

Whether it's a list, flow chart or diagram, the method is the same. Suppose you are trying to memorise a list. This is what you should do:

- Place the list in front of you with a sheet of blank paper beside it.

- Copy out the key facts in the list, making a real effort to remember each one. Copying out is an essential part of learning. It makes it an active process, and the very act of writing or drawing helps to fix the facts in your long-term memory. This is where so many people go wrong; they just stare passively at their notes in the hope that it will sink in. It won't!

   DON'T JUST SIT THERE, DO SOMETHING!

- Put your first list to one side. Take another sheet of paper and copy out the list again, but this time try to do as much as you can without looking at the original.

- Repeat the process until you can write out the list without looking at the original at all.

- **Next day** try writing out the list again, without looking. Unless you have a very good memory, you'll probably have to look at parts of it now and again to refresh your memory. Don't be discouraged! You can't expect it all to go into your long-term memory straight away. Repeat the procedure daily until you can remember the list easily. Then, once a week or so, try writing out the list again. This will reinforce it in your memory. This self-testing is an important part of learning. It teaches you to recall things quickly.

- Shortly before the exam you need to refresh your memory very quickly. Keep all your lists, definitions, flow charts, diagrams and summaries so that you can look at them again when you come to this final revision.

How should I plan and do my revision?

Revising for an examination requires organisation and self-discipline. Here are some hints about how to go about it:

- Put aside a regular amount of time on certain days for your physics, and stick to it.
- Plan how you're going to spend the time. Be sure to spend part of the time testing yourself on work covered in previous sessions.
- Sit in the same place each time, preferably at a desk in a room of your own.
- Keep your desk tidy at all times; you can't work effectively in a mess.
- Some people like listening to music while they are working. If you do find it helps, keep it fairly quiet and listen to tapes rather than the radio to avoid DJs interrupting your train of thought.
- Don't day-dream. Day-dreaming is most tempting when you're tired so don't go to too many late-night parties!
- Avoid being distracted. Don't look out of the window too often and don't allow visitors into your room.
- Have a short break at least once an hour. Working for too long without a break is as bad as having too many breaks. You will also find that after a short break you can concentrate better and think more clearly.
- Keep physically fit. If you don't play a sport regularly, go for a brisk walk every day.
- When you do take a break, let it be a complete break: forget all about GCSE!

QuickCheck Syllabus Grid

CHAPTER	TOPIC	LEAG (A)	LEAG (B)	MEG	MEG (Nuff)	NEA (A)	NEA (B)	NISEC	SEB (standard)	SEG	SEG (alt)	WJEC
1	Measuring	●	●	●	●	●	●	●	●	●	●	●
2	Discovering forces	●	○	●	○	●	○	○	○	●	○	●
3	More about forces	●	○	○	○		○	●	○	●	○	
4	Pressure	○		○	○	○	○	○	○	○	○	●
5	Air pressure	○		○	○		○	○		○	○	
6	Work and machines	○	○	●	○	●	○	●	●	●	○	
7	Simple machines	●	○		○	●	●	●	○	●		
8	Speed	●	●	●	●	●	●	●	●	●	●	●
9	Velocity and acceleration	○	○	●	○	○	●	○	○	●	●	●
10	Newton's laws of motion	●	●	○	●	●	●	●	●	●	●	●
11	Kinetic energy and momentum	○	●	○	●	○	●	●	●	○	●	●
12	Transport	○	●	○	○	○	○	●	●	○	○	○
13	Flight and orbits				●				●			
14	The solar system				●				○			
15	Refraction and colour	○	○	○	○	○	○	○	○	●		●
16	Reflection	●	○	○	○	●	●	○	●	○	○	○
17	Lenses	○	○	○	○	○	●	○	●			
18	Optical instruments	○	○		○	○	●		●		○	
19	Waves	○	○	○	●	○	○	○	○	○	○	●
20	Watching waves	●	●	●	●	○	●	○	●	●	○	●
21	Sound	●	●	●	●	●	○	●	○	○	○	●
22	The electromagnetic spectrum	○	○	●	●	●	○				○	●
23	Static electricity	●			○	●	●				○	●
24	Electric circuits	●	●	●	●	●	●	●	●	●	●	●
25	Resistance	●	●	●	●	●	○	●	●	●	●	●

26 Magnetism
27 Effects of electricity
28 Electricity at home
29 Motors and dynamos

30 Cathode ray tubes
31 Circuit devices and sensors
32 Transistor circuits
33 The operational amplifier

34 Digital electronics
35 Communication systems
36 Materials

37 Heat and temperature
38 Change of state
39 Gases
40 Heat energy gets around

41 People need energy
42 Energy sources
43 Energy conversions

44 Electricity distribution
45 The structure of atoms
46 Radioactivity
47 Nuclear energy
48 Physics and health

Key ● means all of topic, ○ means some of topic.

Measuring

Almost all the units used in physics belong to the SI or International System of units. A list of common units is given on page 158. All basic units can be made bigger or smaller. You do this by adding a metric prefix as shown in the table below.

Metric prefixes.

Name	Symbol	Value	(= × by)
giga	G	10^9	= 1000 000 000
mega	M	10^6	= 1000 000
kilo	k	10^3	= 1000
no prefix		1	
deci	d	10^{-1}	= 1/10
centi	c	10^{-2}	= 1/100
milli	m	10^{-3}	= 1/1000
micro	μ	10^{-6}	= 1/1000 000
nano	n	10^{-9}	= 1/1000 000 000
pico	p	10^{-12}	= 1/1000 000 000 000

How do we measure length and distance?

● The SI unit of **length** and **distance** is the **metre (m)**. We use millimetres for measuring small things. Kilometres are used for distances.

Choice of instrument.

Instrument	Range of lengths	Example
Micrometer screw gauge	0.01 mm to 20 or 30 mm	Thickness of a hair or wire
Ruler	1 mm to 1 m	Height of a table
(Steel) tape measure	100 mm to several metres	A javelin throw

Prefix examples.

Name	Symbol	Value
4 gigawatts	4 GW	4×10^9 watts
2 megajoules	2 MJ	2×10^6 joules
57 kilohertz	57 kHz	57×10^3 hertz
3 decibels	3 dB	0.3 bels
50 centimetres	50 cm	0.5 metres
40 milliamperes	40 mA	0.04 amperes
1 milligram	1 mg	1×10^{-3} grams
8 microamperes	8 μA	8×10^{-6} amperes
5 nanoseconds	5 ns	5×10^{-9} seconds
10 picofarads	10 pF	10×10^{-12} farads

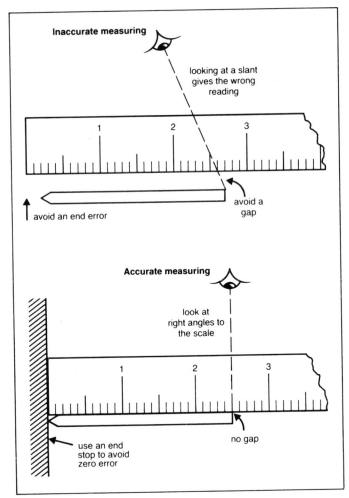

Figure 1.1 *Using a ruler.*

How do we measure area?

● The unit of **area** is the units of length squared.
● The SI unit of area is the **square metre (m^2)**. The square centimetre (cm^2) is often used. Note that $1 m^2 = 100 cm \times 100 cm = 10^4 cm^2$.
Write $1 m^2$ in mm^2, $1 cm^2$ in m^2 and $1 mm^2$ in m^2.

How do we measure volume?

The volume of an object is a measure of the amount of space it occupies.

● The unit of volume is the units of length cubed.
● The SI unit of volume is the **cubic metre (m^3)**. The cubic centimetre (cm^3) is also often used. $1 m^3 = 100 cm \times 100 cm \times 100 cm = 10^6 cm^3$.
Write $1 m^3$ in mm^3, $1 mm^3$ in m^3 and $1 cm^3$ in m^3.

In figure 1.3, the solid object displaces some water from the can. This is collected in the measuring cylinder. The volume equals the volume of the object. (Read the volume of water at the bottom of the meniscus.)

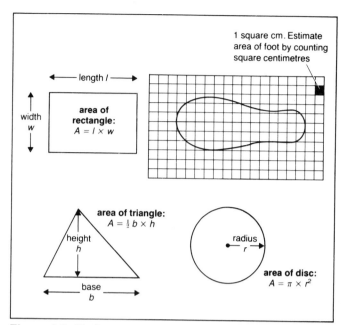

Figure 1.2 *Finding areas.*

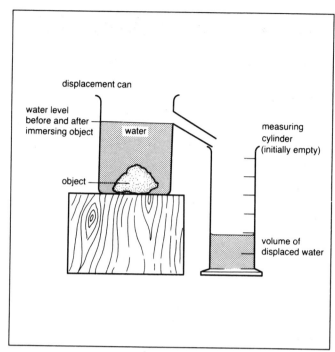

Figure 1.3 *Measuring volumes:*

What is the mass of an object?

The mass of an object is a measure of the amount of matter or 'stuff' there is in it.

We might describe an elephant or a large stone pillar as 'massive'. They both contain a lot of matter. So they both have large masses.

- The SI unit of mass is the **kilogram (kg)**.
  $1000\,kg = 1$ tonne.
- The mass of an object is measured by balancing it against known masses. Use a pair of scales or a 'beam balance'.

What is the density of a material?

- The **density** of a material depends on the mass of its atoms and how closely they are packed together.

Density is the mass per unit volume of a material.

In SI, mass is in kg, volume in m^3. So density is in **kilograms per metre cubed (kg/m³)**. But often it is easier to use grams per centimetre cubed (g/cm³).

$$\text{density} = \frac{\text{mass}}{\text{volume}} \qquad d = \frac{m}{V}$$

Example: *Calculating density.* A glass stopper has a volume of $16\,cm^3$ and a mass of $40\,g$. Calculate the density of glass.

$$d = \frac{m}{V} = \frac{40\,g}{16\,cm^3} = 2.5\,g/cm^3$$

How can we measure time?

- The SI unit of **time** is the **second** (symbol **s**).
- A time interval is written as Δt. This is the length of time between the beginning and end of some event.

The Greek letter Δ means 'change of'. Common techniques for measuring time intervals include:
- a stopwatch with pointer (analogue type);
- an electronic digital stopwatch;
- automatic electronic timing using light beams and photocells.

Some other useful clocks.

The caesium atomic clock. It can measure one nanosecond ($1\,ns = 10^{-9}\,s$) quite accurately.
The quartz crystal oscillator. This is used in watches and computers. It keeps accurate time.
Alternating current of mains electricity. This has a frequency of 50 hertz. It can be used to drive clocks and record players at a steady speed.
The swing of a pendulum. This is used in clocks and metronomes. It keeps accurate time. The time of the swing depends on (the square root of) the length of the pendulum.
One turn of the Earth. This takes one day.
Radioactive decay clocks. Carbon-14 decays in the remains of plants and animals. This is used to measure their age.

Discovering forces

A force is a push or a pull which one object applies to another object. A force has both a size (magnitude) and a direction.

On diagrams an arrow points in the direction of the push or pull.

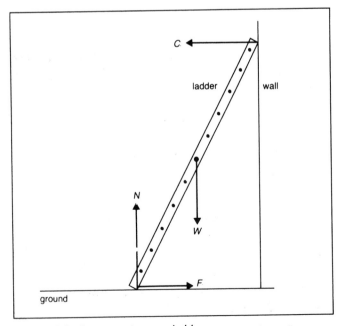

Figure 2.1 *Forces on a parachute:*

R = air resistance or drag. This pushes on the inside of the parachute. It slows down the fall by opposing the weight.
T = tension in the ropes. It pulls up on the man and down on the parachute.
W = weight. This is the pull of gravity on the man.

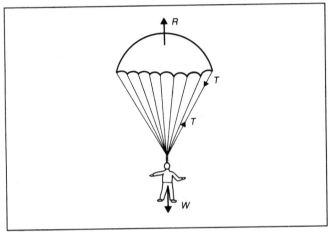

Figure 2.2 *Forces acting on a ladder:*

W = weight of the ladder. This is the downwards pull of gravity on the ladder.
C = contact force of the ladder. This presses against the wall. It causes the wall to push back with an equal and opposite force.
F = friction force. This stops the ladder slipping.
N = normal force. This stops the ladder sinking in.

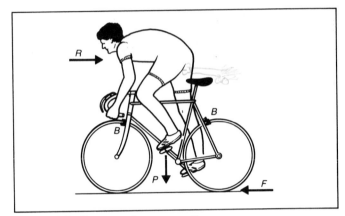

Figure 2.3 *Forces acting on a cyclist:*

R = air resistance. This pushes against the cyclist.
F = friction between the tyre and the road. It is needed for grip and forward acceleration.
P = push of the foot on the pedal. This causes forwards acceleration. It keeps the cyclist going against resisting or frictional forces.
B = brakes. These use friction between brake blocks and wheel rims to slow the cyclist down.

How do we measure forces?

The size or magnitude of a force is measured in **newtons** (symbol **N**). The newton is defined in Chapter 10.
Forces can be measured using a spring balance called a **'newton meter'**. It has a scale labelled in newtons.

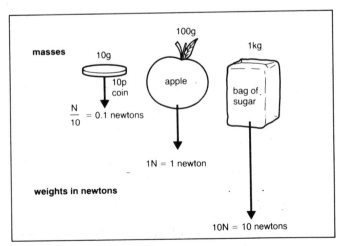

Figure 2.4 *Forces measured in newtons:*

4

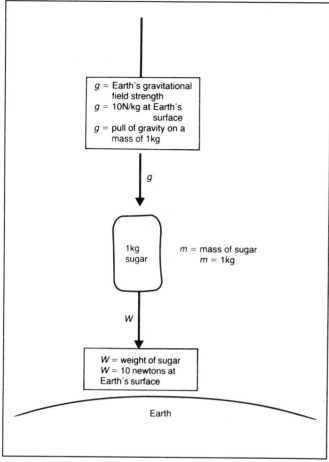

Figure 2.5 *Mass and weight:*

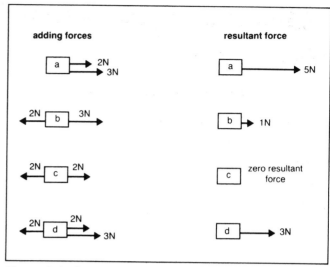

Figure 2.6 *Combining parallel forces*

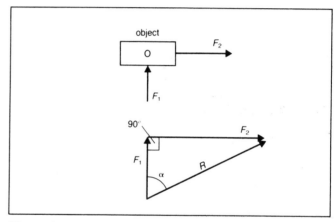

Figure 2.7 *Adding forces at right angles:*

The mass *m* is a scalar. Weight *W* is a vector.
$W = m\,g$ where *m* is in kg, *W* is in N and *g* is in N/kg.
Mass is the same everywhere.

How are vectors and scalars different?

A scalar is a quantity which has size but no direction.

● Some scalars are: temperature, energy, power, density, mass and volume.
● Scalars can be simply added together.
For example, volumes: $5\,cm^3 + 4\,cm^3 = 9\,cm^3$.

A vector is a quantity which has both size and direction.

● Some vectors are: force, velocity, acceleration and momentum.
● When vectors (such as forces) are added their directions must be taken into account.

Object O is being pushed by force F_1 and pulled by force F_2. When F_1 and F_2 are drawn to scale the size of the resultant *R* is given by its length.
The size of *R* is given by:

$$R^2 = F_1^2 + F_2^2$$

The direction of *R* is at an angle α to F_1.

$$\text{Tan } \alpha = \frac{F_2.}{F_1}$$

Use inverse or arc tan on your calculator to find angle α.

Example: $F_1 = 3\,N$. $F_2 = 4\,N$. F_1 and F_2 are at 90°.
Find *R* and α.
$R^2 = 3^2 + 4^2 = 9 + 16 = 25$. So $R = 5$ newtons.
Tan $\alpha = 4/3$ which gives $\alpha = 53°$.

3 More about forces

What is the moment of a force?

We use forces to open doors, tighten nuts and pedal bicycles. In these cases the forces have a *turning effect* on the objects.

The turning effect of a force is called the moment of the force.

Think of two more examples of forces which have a turning effect or moment.

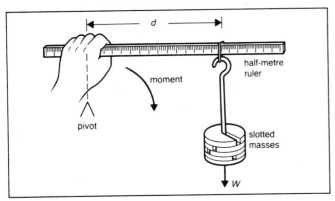

Figure 3.1 *Feeling the moment of a force:*

The experimenter feels a greater moment (twisting or turning effect) if:

- the weight W of the masses is increased;
- the distance d of the masses from the pivot is increased.

The moment of a force depends on both the size of the force and how far away it is from the pivot.

The pivot or turning point is also called the **fulcrum**. It can be a knife edge, an axle, a hinge or the corner of an object.

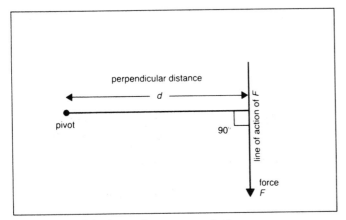

Figure 3.2 *Calculating the moment of a force:*

- The moment M of a force is:

  the magnitude of the force F $\times$ the perpendicular distance d from the pivot to the line of action of the force

$$M = F \times d$$

- F and d are at right angles.
- If F is in newtons and d in metres, the moment M is in **newton metres (N m)**.

Example: *Calculating the moment of a force.* The crank of a bicycle pedal is 16 cm long. The downwards push of a leg is 400 N. Calculate the moment of the push when the crank is horizontal. What difference will it make to the moment if the crank is not horizontal?

$$\text{moment } M = F \times d$$
$$\therefore \quad M = 400\,\text{N} \times 0.16\,\text{m} = 64\,\text{N m}$$

When the crank turns a little further round, the perpendicular distance from the pivot to the line of action of the push gets shorter. So the moment gets smaller. When the pedal is at the bottom the perpendicular distance is zero. So the moment is zero.

What is a centre of gravity or centre of mass?

The centre of gravity (or centre of mass) of an object is the point on it where gravity seems to act.

- The whole weight of an object seems to pull down at its centre of gravity.

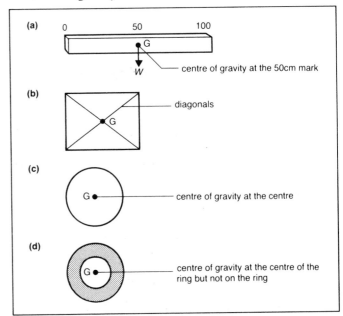

Figure 3.3 *Centres of gravity of regular shapes:*

(a) a metre ruler of uniform thickness. (b) A rectangular-shaped piece of card or plywood of uniform thickness. (c) A disc of uniform thickness. (d) A ring of uniform thickness.

When do objects balance?

Objects *balance* when the moments of the forces acting on them are balanced.

Forces which tend to turn an object in a clockwise direction have a clockwise moment. If they turn an object the opposite way they have an anticlockwise moment.

The law of moments or principle of moments

When an object is balanced:

the *sum of the anticlockwise moments about any point = the sum of the clockwise moments about the same point

(*'sum of the moments' means 'add the moments turning the same way but do not add the forces'.)

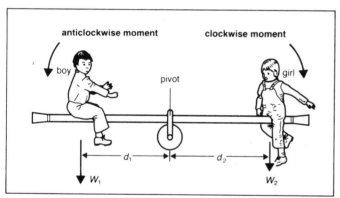

Figure 3.4

The seesaw will balance when:

anticlockwise moment of the boy's weight = clockwise moment of the girl's weight
$$W_1 \times d_1 = W_2 \times d_2$$

Example: *Using the law of moments.* A boy of weight 500N sits on the left side of a seesaw. He is 2.4 metres from its pivot. A girl can balance the seesaw by sitting 3.0 metres from the pivot on the right side. What is her weight?

Refer to the diagram above. Using the law of moments, when the seesaw is balanced:

anticlockwise moment of the boy's weight = clockwise moment of the girl's weight
$$W_1 d_1 = W_2 d_2$$
$$500 \text{ N} \times 2.4 \text{ m} = W_2 \times 3.0 \text{ m}$$
$$\therefore \text{ the girl's weight } W_2 = \frac{1200 \text{ N m}}{3.0 \text{ m}} = 400 \text{ N}$$

When will an object be in equilibrium?

A balanced object is said to be in equilibrium.

An object will be in equilibrium if:

● the forces acting on it are balanced (i.e. there is no resultant force);
● the moments of the forces are balanced.

When is an object stable?

An object is said to be stable or has stable equilibrium if it returns to its upright position after being knocked or slightly tilted.

This happens if its centre of gravity *rises* when it is tilted.

An object can be made more stable (less likely to topple over) if you:
● lower its centre of mass;
● fit a wider base.

A *hanging* object has its centre of gravity *below* its support, so it is always stable.

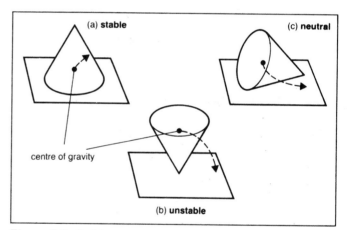

Figure 3.5 *Stability:*

(a) A cone in *stable equilibrium* is hard to tip over. Tilt it and its centre of gravity rises. Let it go and it falls back.

(b) A cone in *unstable equilibrium* falls over when it is tilted. Its centre of gravity falls.

(c) A cone in *neutral equilibrium* rolls. It keeps its centre of gravity at a constant height.

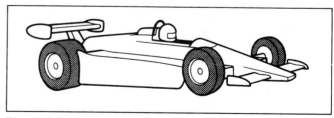

Figure 3.6 *A car designed to be very stable:*

It has a wide wheel base and a low centre of gravity.

Review questions: Chapters 1, 2 and 3

C1 1 Copy and complete the following:
  (a) An object's *volume* is a measure of _____.
  (b) An object's *mass* is a measure of _____.
  (c) Two factors which determine the density of a material are _____ and _____.
  (d) The *density* of a material is _____.
  (e) A quartz crystal oscillator, mains electricity and a pendulum swing can all be used as _____.

2 Give the SI units of the following:
  (a) length;   (b) time;   (c) volume;   (d) mass;
  (e) distance;   (f) density;   (g) area.

3 Name a suitable instrument to measure:
  (a) a discus throw;
  (b) the diameter of a piece of fuse wire;
  (c) the length of a textbook;
  (d) the mass of a bag of potatoes;
  (e) the diameter of a marble.

4 Give the symbol and the value of the following metric prefixes:
  (a) milli;   (b) micro;   (c) kilo;   (d) mega;
  (e) centi;   (f) nano;   (g) giga;   (h) pico.

5 How many millimetres are there in:
  (a) 1 cm; (b) 4 cm; (c) 50 cm; (d) 1 m; (e) 1.2 m?

6 Write the following without using prefixes:
  (a) 4 MW; (b) 3 mm; (c) 1 μA; (d) 6 cm;
  (e) 7.3 kHz.

7 Rewrite using the most suitable prefix:
  (a) 5000 joules;
  (b) 0.02 metres;
  (c) 4/1000 (or 0.004) grams;
  (d) 0.000 001 seconds.

8 Write: (a) 2 m^2 in mm^2;   (b) 4 m^2 in cm^2;
  (c) 3 mm^2 in m^2;   (d) 1 mm^3 in m^3;
  (e) 8 cm^3 in m^3;   (f) 5 m^3 in mm^3.

9 A block of perspex has a volume of 40 cm^3 and a mass of 48 g. What is the density of perspex?

10 A silver ring has a volume of 0.5 cm^3. The density of silver is 10.5 g/cm^3. What is the mass of the ring?

C2 11 Copy the diagrams shown. Add labelled arrows to show the forces acting on the objects:

  (a) A helicopter hovering.
  (b) A car moving forward in a straight line.
  (c) A bowl of flowers on a table.
  (d) A plank of wood leaning against a wall.

12 Copy and complete the following:
  (a) A *vector* has _____ and _____.
  (b) A *scalar* has _____ but no _____.
  (c) The SI unit of *force* is _____.
  (d) On force diagrams an _____ points in the direction of the push or pull.
  (e) The symbol for the *gravitational field strength* of the Earth is _____, units _____.

13 Which of the following are *vectors*:
  (a) mass; (b) weight; (c) velocity; (d) speed;
  (e) temperature; (f) density; (g) momentum?

14 A rescue helicopter lifts a girl out of the sea. Her mass is 50 kg. What is the tension in the rope due to the girl's weight? (Assume $g = 10 \text{ N/kg}$.)

15 In the formula $W = m\,g$, what do the letters stand for? What are their SI units?

On Earth g has a value of 10 in SI units.

On Pluto g has a value of 4 in SI units.

Assume identical statues of 20 kg are found on Earth and on Pluto.
(a) What are the values of m for the statues on Earth and on Pluto?
(b) What are the values of W for the statues on Earth and on Pluto?

16 A shopping bag has a mass of 0.2 kg. It contains a 1 kg bag of sugar, a 0.4 kg tin of peas and 0.4 kg of ham. What is:
(a) the mass of the bag and purchases?
(b) the total weight of the load?
(c) the resultant force the load exerts on the shopper's hand? (Take $g = 10$ N/kg.)

17 Alex and Anne are pulling a boat out of a river. They both pull in the same direction. Their forces are 100 N and 80 N respectively. What is their total force?

The friction on the boat is 30 N. What is the resultant force on the boat?

18 Two boys are trying to move a stone. Tom pulls with a force of 50 N. Tim pushes with a force of 60 N. This is shown in the figure.

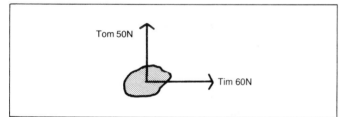

(a) What is the total force on the stone?
(b) Which way will it move?

19 Copy and complete the following:
(a) The *moment* of a force is a _____ effect.
(b) The *pivot* is also called the _____.
(c) The *moment* of a force can be calculated by multiplying the force by the _____.
(d) The *angle* between the force and the distance must equal _____.

20 Susan opens a door by pulling the handle with a force of 60 N.
(a) Where is the fulcrum?
(b) If the fulcrum is 75 cm from the handle, calculate the moment of the force.

21 Sketch diagrams of the following. Indicate the approximate position of their *centre of gravity*:
(a) a hanging basket of flowers;
(b) a car tyre;
(c) a floorboard;
(d) a sugar lump;
(e) an empty ice-cream cornet.

22 A boy of weight 400 N sits on the left side of a seesaw. He is 2.0 m from the pivot. A girl can balance the seesaw by sitting on the other side. She is 2.5 m from the pivot as shown in the figure.

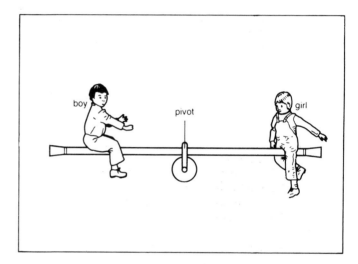

(a) What is the size of the anticlockwise moment?
(b) Assume the seesaw is balanced. What must be the size of the clockwise moment?
(c) What is the weight of the girl?

23 Copy and complete the following:
(a) An object is stable if, when it is knocked, it _____.
(b) By lowering its centre of mass an object becomes _____ stable.
(c) By fitting a smaller base an object becomes _____ stable.
(d) A hanging object will be stable if its centre of gravity lies _____ its support.

24 Draw diagrams of a pear in:
(a) stable equilibrium;
(b) unstable equilibrium;
(c) neutral equilibrium.

25 Write some notes on the stability of:
(a) hyacinths in soil in a low bowl compared to tall daffodils in a thin vase;
(b) the crew leaning out over the side of a yacht in a race.

Pressure

How is pressure related to force?

Pressure is a measure of how a force acting on an object is spread out over its surface.

> **Pressure is defined as the normal force acting on a unit of surface area.**

A '*normal*' force is one which acts 'at right angles to' the surface.

The unit of pressure equals the unit of force divided by the unit of area.

In SI units it is **newtons per square metre** (N/m^2).

1 newton per square metre is called a **pascal** (**Pa**).

$$\text{pressure} = \frac{\text{normal force}}{\text{area}}$$

$$p = \frac{F}{A} \qquad \boxed{\begin{array}{c} F \\ \hline p \times A \end{array}}$$

Example: *Calculating pressure.* A large crate has sides of length 2 m, 5 m and 8 m. The crate and its contents weigh 600 N. What is the pressure under it when it is laid down flat? What is the pressure when it is standing on end?

flat	on end
$\text{pressure} = \dfrac{F}{A}$	$\text{pressure} = \dfrac{F}{A}$
$p = \dfrac{600\,N}{5\,m \times 8\,m} = \dfrac{600\,N}{40\,m^2}$	$p = \dfrac{600\,N}{5\,m \times 2\,m} = \dfrac{600\,N}{10\,m^2}$
$= 15\,N/m^2$	$= 60\,N/m^2$

There is a higher pressure under the smaller area.

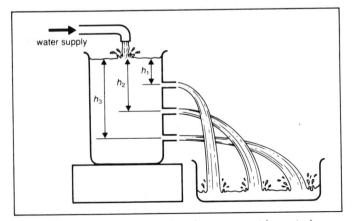

Figure 4.1 *The pressure in a liquid increases with vertical depth below the surface:*

The height of water above (i.e. the depth) increases from h_1 to h_2 to h_3. The pressure also increases. This forces the water out faster and further.

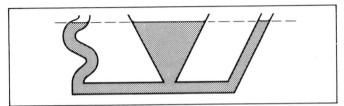

Figure 4.2 *Liquid surfaces all at the same level:*

The pressures at the bottom of all the tubes are the same. This is because there is the same vertical height of liquid above. Liquid flows from one tube to another only when there is a pressure difference.

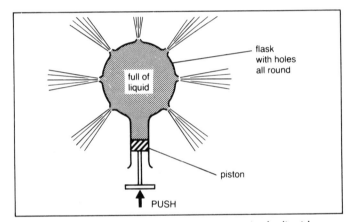

Figure 4.3 *Pushing the piston puts pressure in the liquid:*

The pressure, acting equally in all directions, forces liquid out equally in all directions.

What affects the pressure inside a liquid?

Pressure in a liquid.

Pressure depends on	Pressure does not depend on
Height h of liquid above Density d of liquid	Shape of container Direction – it is equal in all directions

How do we calculate liquid pressure?

$$\text{liquid pressure} = \text{height} \times \text{density} \times g = h\,d\,g$$

In SI units h is in metres, d is in kg/m^3 and $g = 10\,N/kg$. Liquid pressure is then in N/m^2 or pascals (Pa).

Example: The density of mercury is $14000\,\text{kg/m}^3$. A column of mercury in a barometer is $0.75\,\text{m}$ high. Calculate the pressure at the base of the mercury column.

$$p = h\,d\,g = 0.75\,\text{m} \times 14000\,\text{kg/m}^3 \times 10\,\text{N/kg}$$
$$\therefore \quad \text{pressure} = 105000\,\text{N/m}^2 \text{ or } 105\,\text{kPa}$$

How do hydraulic machines work?

Hydraulic machines use liquid pressure to:

- multiply forces;
- transmit forces from one place to another.

Hydraulic machines work because:

- liquids are (almost) incompressible;
- the pressure in a liquid acts equally in all directions;
- pressure changes are transmitted instantaneously through a liquid.

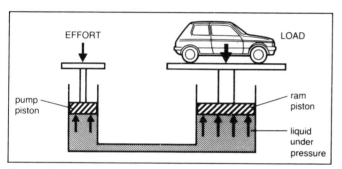

Figure 4.4 *The hydraulic jack:*

$$\frac{\text{Liquid pressure}}{\text{from pump piston}} = \frac{\text{effort}}{\text{area of pump piston}}$$

$$\frac{\text{upwards force acting}}{\text{on ram piston}} = \frac{\text{liquid}}{\text{pressure}} \times \frac{\text{area of}}{\text{ram piston}}$$

$$\frac{\text{force on}}{\text{ram piston}} = \text{effort} \times \frac{\text{area of ram piston}}{\text{area of pump piston}}$$

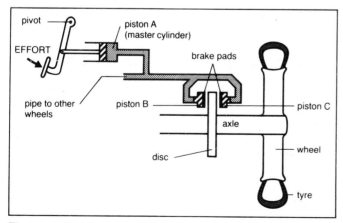

Figure 4.5 *Hydraulic brakes:*

Example: A hydraulic jack is used to lift a car. It has a pump piston of area $2.0\,\text{cm}^2$. The ram piston has an area of $200\,\text{cm}^2$. The effort applied to the jack is $120\,\text{N}$. What force will be applied to the ram to lift the car?

$$\text{force on ram} = 120\,\text{N} \times \frac{200\,\text{cm}^2}{2.0\,\text{cm}^2} = 12000\,\text{N}$$

This would lift a car of mass $1200\,\text{kg}$ or 1.2 tonnes.

Pushing the pump piston A makes the pressure rise everywhere in the brake fluid. The fluid pressure then pushes the ram pistons B and C out. These press the brake pads against the discs and slow down the car.

Why do some objects float?

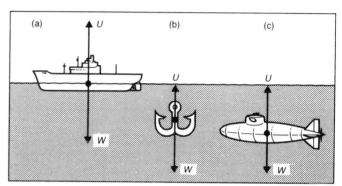

Figure 4.6 *Sinking and floating:*

Two vertical forces act on an object immersed in water. They are the weight W, and the **upthrust** or **buoyancy force** U.

In (a) a floating ship, weight equals upthrust ($W = U$).

In (b) a sinking anchor, weight is greater than upthrust ($W > U$).

In (c) a rising submarine, weight is less than upthrust ($W < U$).

What did Archimedes discover?

Archimedes discovered that the upthrust was always equal to the weight of fluid displaced by the object. ('Displaced' = 'pushed out of the way'.)

The law of flotation

An object will float if it can displace a weight of fluid equal to its own weight.

- A steel ship can float, even though steel is denser than water. This is because it displaces a weight of water equal to its own weight.
- A submarine has flotation tanks. It can fill these with either air or water. The weight of the submarine is increased by filling them with water. This makes it sink. When they are filled with air the submarine rises.

Air pressure

Effects and uses of air pressure

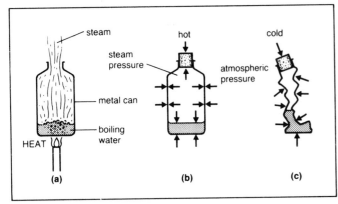

Figure 5.1 *The can-crushing experiment:*

(a) All the air is driven out by filling the can with steam.
(b) While hot, the steam pressure inside balances the air pressure outside.
(c) When cold, steam condenses so that the inside pressure falls. The much greater air pressure outside crushes the can.

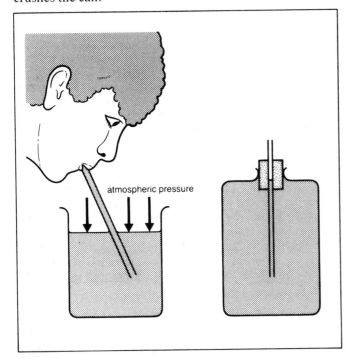

Figure 5.2 *Sucking up a straw:*

Sucking up a straw is done by reducing the pressure inside your mouth so that air pressure on the surface of the drink pushes it up the straw.

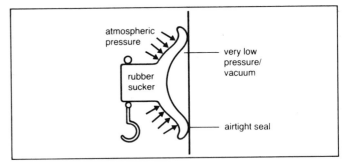

Figure 5.3 *Rubber suckers are held on to a surface by the air pressure on the sucker:*

Pressing a moist sucker against a smooth surface squeezes out the air. The moisture (or grease) seal keeps the pressure behind the sucker lower than the air pressure. Industry uses suckers to lift large sheets of glass and metal.

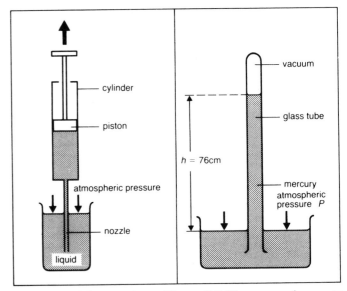

Figure 5.4 *Filling a syringe:* **Figure 5.5** *The mercury barometer:*

The piston in a syringe is raised. This reduces the pressure below the piston. The air pressure then pushes liquid up into the space.

How do we measure pressure?

The air pressure is $P = hdg$. h is the height of the mercury column and d is its density. (However, air pressure is often quoted as the height of a mercury column in mm.)

Example: The mercury column is 750 mm high. The density of mercury is 14 000 kg/m³. What is the air pressure in pascals? (In SI units 750 mm = 0.75 m.)

$$\text{pressure} = hdg = 0.75\,\text{m} \times 14\,000\,\text{kg/m}^3 \times 10\,\text{N/kg}$$
$$\therefore \quad \text{air pressure} = 105\,000\,\text{Pa or } 105\,\text{kPa}$$

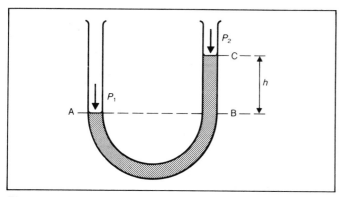

Figure 5.6 *The U-tube manometer measures how much greater one pressure is than another:*

The pressure difference is shown by the difference in liquid levels h. Density of the liquid is given by d.

The pressures at A and B are equal because they are at the same level in the liquid.

P_1 is the pressure to be measured (say the gas pressure).

Pressure at B = air pressure P_2 + pressure due to liquid column CB. So the gas pressure $P_1 = P_2 + h d g$.

The gas pressure is greater than the air pressure by $h d g$.

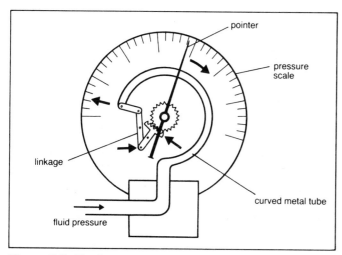

Figure 5.7 *The Bourdon gauge:*

This type of pressure gauge is used on gas cylinders. It measures fluid (i.e. gas or liquid) pressure. The tube tries to uncurl or straighten out when the pressure inside the curved metal tube increases. (The linkage magnifies the tube's movement.)

The aneroid barometer

Another type of barometer uses a flat, partially evacuated, cylindrical metal box. It is known as an **aneroid barometer** and is used as an altimeter in aircraft. The scale is calibrated in height above sea level based on the fall in atmospheric pressure as the aircraft flies higher.

What is atmospheric pressure?

The atmospheric or air pressure at sea level is about 100 kPa. This is taken to be 'normal' atmospheric pressure. It is called 1 **bar**.

On the weather maps which we see on television, the pressures are given in **millibars**.

1 bar = 1000 millibars = 760 mm of mercury

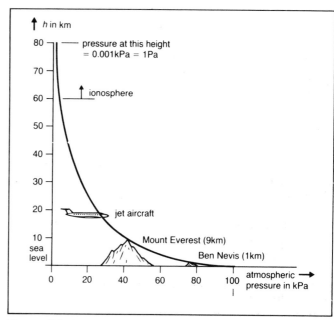

Figure 5.8 *Variation of air pressure with height above sea level:*

The air pressure at the ground is caused by the weight of all the air above it. Suppose we go higher in the atmosphere, for instance when in an aircraft or climbing a mountain. Then there is less air above to compress the air here. So the pressure is lower.

How does pressure affect people?

The change in air pressure when we gain or lose height quickly causes our ears to 'pop'. The ear drum, which separates the middle ear from the outside air, bulges out when the outside air pressure falls below that of the air trapped inside the ear. Swallowing opens a tube connecting the middle ear to the throat, so allowing the air pressure inside the ear to equalise with that of the outside air.

Explain why the same effect is experienced by travellers on a high speed train when it enters a tunnel.

It is necessary to pressurise the cabins of high-flying aircraft because the air pressure is so low that people could not breathe enough oxygen. Astronauts above the atmosphere live in pressurised cabins or wear pressure suits which, as well as supplying the necessary oxygen, prevent the blood and water in their bodies from boiling.

6

Work and machines

How much work is done by a force?

● **Work** is done only when a force *moves* an object in the *same* direction as the force.

If you stand holding a heavy parcel or you try pushing a wall over, you do no work since nothing moves.

● Work done = force × distance moved in the direction of the force

$$W = F \times s$$

● If F is in newtons (N) and s is in metres (m), then the work done W is in newton metres (N m).
● The SI unit of work is the **joule (J)**.
● 1 joule = 1 newton metre.

Example: A brick of mass 2 kilograms is lifted through a vertical height of 2 metres. Find the work done.
To lift the brick a height s an upwards force F (equal and opposite to its weight) is needed.

weight of brick = $m\,g$ = 2 kg × 10 N/kg = 20 N
work = $F\,s$ = 20 N × 2 m = 40 N m or 40 joules

Does a machine use up energy?

Fuel (or the energy stored in fuel) gives a machine **the ability to do work.**
    A machine does work when:
● it applies a force to move a load at one place, using another force (the effort) input at a different place (e.g. when an effort is applied to one end of a lever, the other end of the lever applies a force to move the load);
● it converts the energy supplied to it into another form (e.g. a car engine converts the energy stored in petrol into kinetic (or motion) energy and heat energy).

Machines do not 'use up' or 'consume' energy even though they need it to be able to work. Energy is **conserved** because the total energy output equals the energy input.

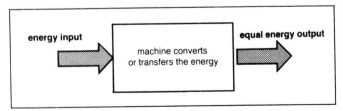

Figure 6.1 *Energy is conserved in a machine:*

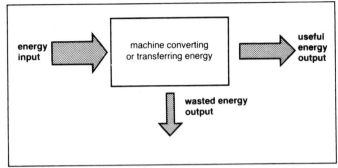

Figure 6.2 *The machine also does work against frictional forces:*

This work wastes input energy by converting it to useless heat. Here the total energy output is the sum of the useful energy output and the wasted energy output. (The energy input is calculated from: effort × distance the effort moves. The useful energy output is calculated from: load × distance the load moves.)

Why use a machine?

① To multiply forces: to allow heavier loads to be moved.

② To multiply distances: to convert small movements into larger ones, to do work faster or to move faster.

③ To do work where it is unsafe or unpleasant for people to go.

④ To do a job more accurately than people can.

⑤ To convert energy from one form into another.

How efficient is a machine?

$$\text{efficiency} = \frac{\textbf{useful work output}}{\textbf{work input}} \text{ or } \frac{\textbf{useful energy output}}{\textbf{energy input}}$$

The **efficiency** of a machine, being a ratio, has no units. Its value will always be less than 1. To express efficiency as a percentage, multiply by 100.

Example: *Efficiency of a pulley system.* A man uses a pulley system to lift a car engine of weight 2000 N a height of 1 metre. He pulls the rope a distance of 8 m. A force of 300 N is used.

useful work output = $F\,s$ = 2000 N × 1 m = 2000 N m
work input   = $F\,s$ = 300 N × 8 m = 2400 N m

∴ efficiency of pulley system = $\dfrac{2000\,\text{N m}}{2400\,\text{N m}}$ = 0.83 or 83%

How powerful is a machine?

The power *P* of a machine is a measure of how much work *W* it does in a certain time *t*.

the power of a machine = its rate of doing work

the power of a machine	=	the rate of conversion of energy from one form to another

$$\text{power} = \frac{\text{work}}{\text{time}} = \frac{\text{energy converted}}{\text{time}} \qquad P = \frac{W}{t}$$

work done or energy converted $W = P \times t$

$$\boxed{\begin{array}{c} W \\ \hline P \times t \end{array}}$$

- The unit of power is the **watt (W)**.
- 1 watt is a rate of doing work of 1 joule per second.
- 1 watt = 1 joule per second (1 W = 1 J/s).

Example: In 4 seconds a girl runs upstairs, a vertical height of 3 metres, lifting her own weight of 440 newtons. The work she does is given by $W = Fs$:

$$W = Fs = 440\,\text{N} \times 3\,\text{m} = 1320\,\text{J}$$

Her power $P = \dfrac{W}{t} = \dfrac{1320\,\text{J}}{4\,\text{s}} = 330\,\dfrac{\text{J}}{\text{s}}$ or 330 watts

---

Example: A lawn mower has an electric motor of input power 800 watts. How much electrical energy does it convert in 10 minutes?

$$\text{energy converted } W = Pt = 800\,\text{W} \times 600\,\text{s}$$
$$W = 480\,000\,\text{J or } 480\,\text{kJ}$$

The efficiency of a machine

$$\text{efficiency} = \frac{\text{useful power output}}{\text{power input}}$$

For example, suppose the rate of working of the electric motor above was measured to be 600 watts (useful output power) when its power intake was 800 watts. Its efficiency would then be 0.75 or 75%.

What is potential energy?

Potential energy is stored energy waiting to do work.

Gravitational potential energy

When the stored energy is caused by gravity and depends on the position or height of an object, it is called **gravitational potential energy**.

Examples:
- A grandfather clock is driven by heavy weights which slowly fall, giving up potential energy.
- The water in a reservoir above a hydroelectric power station has potential energy which is used to drive the turbines and produce electrical energy.

Strain potential energy

When the stored energy which depends on the *state* of *condition* or an object which may be bent, twisted, stretched or compressed, it is called **strain potential energy**.

Examples
- A clockwork motor is driven by the strain potential energy stored in a coiled spring.
- An arrow is fired using the strain potential energy stored in a bent bow.

How do we calculate potential energy, *E*p?

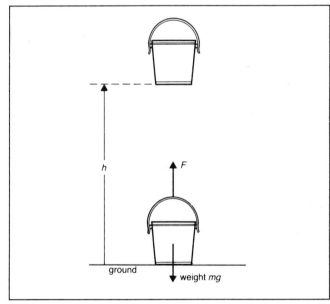

Figure 6.3 *Calculating potential energy:*

The gravitational potential energy E_p possessed by an object of mass m is equal to the work done in lifting it to a vertical height h above the ground. A force equal and opposite to the weight of an object is needed to lift it: $F = mg$.

$$\text{work done} = F \times h = mgh$$
$$\text{potential energy } E_p = mgh$$

Example: The weight in a grandfather clock has a mass of 2 kg. It can fall 0.8 m. How much gravitational potential energy can it store?

$$E_p = mgh = 2 \times 10 \times 0.8 = 80 \text{ J}$$

7

Simple machines

What are force multipliers?

Machines which allow a small effort to move a larger load are called **force multipliers**.

The mechanical advantage (MA) of a machine is the number of times that the load moved is greater than the effort used.

$$\text{Mechanical Advantage} \quad MA = \frac{\text{load}}{\text{effort}}$$

- MA is a ratio and has no units.
- Force multipliers have a MA *greater* than 1 (MA > 1).

Examples of force multipliers

crowbar, wheelbarrow, nutcracker, bottle opener.

What are distance multipliers?

Machines which are designed as **speed** or **distance multipliers** take a small movement of the effort and multiply it to produce a larger movement of the load.

$$\text{Velocity Ratio} \quad VR = \frac{\text{distance moved by the effort}}{\text{distance moved by the load}}$$

- VR is also a ratio and has no units.
- Distance multipliers have a VR *less* than 1 (VR < 1).

Examples of distance multipliers

Human forearm, bicycle, fishing rod.
- A simple machine cannot be both a force and a distance multiplier.
- The VR of a machine is always greater than its MA.

What do levers do?

Levers are simple machines which use a pivot or fulcrum to transfer the work done by the effort at one point to a load at another point.

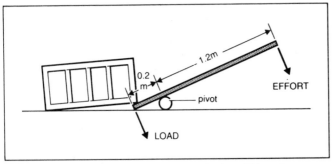

Figure 7.1 *The crowbar is a force multiplier, MA > 1:*

The effort has a greater moment (turning effect or 'leverage') than the load. This is because the perpendicular distance of the effort from the pivot is greater than that of the load.

Example: Assume the distances are as shown in figure 7.1. The crate has a weight of 2400N. Find the effort needed to lift the crate.

$$\text{clockwise moment} = \text{anticlockwise moment}$$
$$\text{effort} \times 1.2\,\text{m} = 2400\,\text{N} \times 0.2\,\text{m}$$
$$\therefore \quad \text{effort} = \frac{2400\,\text{N} \times 0.2\,\text{m}}{1.2\,\text{m}} = 400\,\text{N}$$

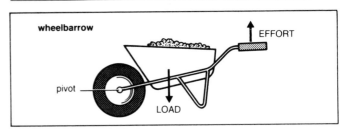

Figure 7.2 *The wheelbarrow:*

The wheelbarrow uses a lever to lift a load. Here the effort is further from the pivot than the load. So the wheelbarrow is a force multiplier with MA > 1. The effort needed is smaller than the load lifted.

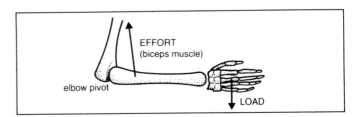

Figure 7.3 *The forearm:*

The forearm acts as a lever to lift an object such as a bucket. When the biceps moves a short distance the load is raised a larger distance. So this lever is a distance multiplier with VR < 1.

What do pulleys do?

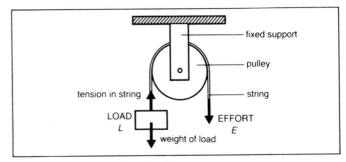

Figure 7.4 *A single fixed pulley:*

This pulley does not move with either the effort or the load. Its position is *fixed*.

A **fixed pulley** is used to convert the direction of the effort E from a downwards pull into an upwards lift (VR = 1).

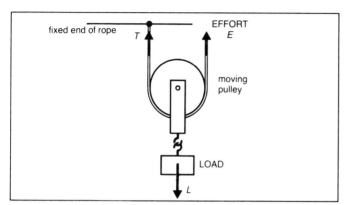

Figure 7.5 *A single moving pulley:*

For any distance the load is raised there are two equal lengths of rope to be pulled up by the effort. So the effort moves twice as far as the load (VR = 2). As the load L hangs on two ropes the effort E supports only $\frac{1}{2}L$. So a single moving pulley is a force multiplier.

- For $E = \frac{1}{2}L$ we would expect MA = 2. However, the MA of this pulley will be < 2. This is because the effort E has to overcome friction and to lift the pulley as well as the load.
- In general, the VR of a pulley system is given by the number of ropes supporting the lower, moving block of pulleys.

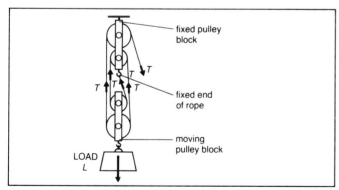

Figure 7.6 *This block and tackle is fitted with two fixed and two moving pulleys:*

The block and tackle is a more effective force multiplier than the single pulley.

- The moving block is supported by four ropes. This gives a VR of 4. The MA is somewhat less than 4.
- The MA can be found only by experimental measurement. It increases towards 4 for larger loads.

What factors make the MA less than 4?
Why does the MA increase as the load increases?

Why use an inclined plane?

An inclined plane is a ramp or slope.

It is easier to push or pull a heavy object up a slope than it is to lift it vertically. So this machine multiplies the effort.

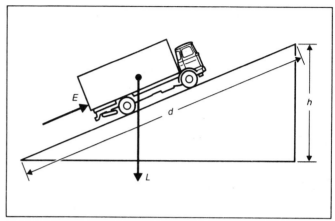

Figure 7.7 *The inclined plane:*

The effort E moves a distance d up the slope. The load L is raised through a vertical height h.

$$VR = \frac{\text{distance effort moves up the slope}}{\text{vertical height load is raised}} = \frac{d}{h}$$

Why is friction needed in machines?

- Grip is needed between a tyre and the road.
- Grip between clutch plates in a car allows smooth gear changes.
- Brakes need it to slow down a vehicle.
- A nut needs friction to stay tight on a bolt.
- In a tape-recorder the tape is gripped and driven by a rubber pinch-wheel.
- In a type-writer the paper is held and moved by a friction grip.
- Things are gripped between finger and thumb using friction.

How is friction reduced in machines?

- Make surfaces which move as smooth as possible.
- Use oil to separate moving surfaces.
- Use wheels or rollers to move heavy objects.
- Use ball or roller bearings in rotating wheels or axles.
- Use a cushion of air to support a vehicle e.g. a hovercraft.
- Use air as an elastic bearing to absorb shocks or vibrations.
- Make machines which move through air or water streamlined in shape.

Review questions: Chapters 4, 5, 6 and 7

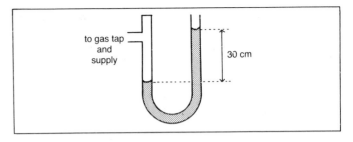

to gas tap and supply

30 cm

C4 **1** Copy and complete the following:
 (a) A force which acts at right angles to an object is called a _____ force.
 (b) The SI unit of pressure is _____.
 (c) Pressure in a liquid _____ with the depth below the surface.
 (d) The pressure in a liquid does not depend upon the _____ of the container.

2 A vase of flowers stands on a birthday cake. It exerts a force of 1 N. The area of the base of the vase is $5 \times 10^{-4} \text{m}^2$.
 Calculate the pressure on the icing.

3 Explain why:
 (a) castor cups are often placed under chairs which stand on carpets;
 (b) a ballet dancer pivoting on one toe exerts more pressure on the floor than if standing on the whole foot.

4 A diver swims in sea water at a depth of 6 m. The density of sea water is 1030kg/m^3. What is the pressure of water on her? (Assume $g = 10 \text{N/kg}$.)

5 A jug contains 10 cm of paraffin oil of density 800kg/m^3. What is the pressure of the oil on the bottom of the jug? (Assume $g = 10 \text{N/kg}$.)

6 A hydraulic jack has a pump piston of area 4cm^2 and a ram piston of area 300cm^2. If the effort applied to the jack is 180 N, what is the force which will be applied to the car?

7 Copy and complete the following:
 (a) An object floats when its weight equals the weight of the _____.
 (b) A stone sinks in water because its _____ is bigger than the _____.
 (c) A canoe will float because it displaces a weight of _____ equal to its own weight.

C5 **8** A U-tube manometer is connected to the gas supply in a laboratory. The mercury in the manometer rises by 30 cm, as in the figure.

 (a) What is the difference between the air pressure and the gas pressure?
 (b) What is the gas pressure? (Assume that the density of mercury is $14\,000 \text{kg/m}^3$, air pressure is 100 kPa and $g = 10 \text{N/kg}$.)

9 A mercury barometer is used to measure atmospheric pressure.
 (a) The weather man says that there is an area of low pressure coming across the country. What changes will occur in the barometer?
 (b) What is the height of the mercury when the air pressure is 'normal'?
 (c) If the height of mercury in the barometer is 765 mm, calculate the atmospheric pressure. (Assume the density of mercury is $14\,000 \text{kg/m}^3$ and $g = 10 \text{N/kg}$.)

10 A syringe consists of a barrel and a plunger.
 (a) How would you partly fill the syringe with water?
 (b) Why does the water go into the syringe?

11 (a) What does an altimeter measure?
 (b) Which instrument is usually used as an altimeter?

C6 **12** Copy and complete the following:
 (a) Work is done when a force moves an object in the same _____ as the force.
 (b) The efficiency of a machine is _____ than 1.
 (c) Power is a measure of the _____ done in time t.
 (d) Stored energy is called _____ energy.

13 A man pushes a pram with a force of 80 N.
 (a) The pram moves 1.5 m as shown in the figure. Find the work done.

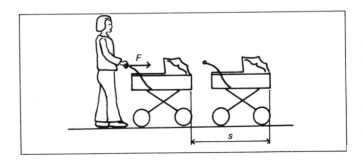

(b) The man lifts the baby from the floor to the pram. How much work does he do? The weight of the baby is 45 N. The vertical height moved is 1 m.

✓

14 A 10 kW electric motor is used to raise a 400 kg load of bricks to a height of 60 m. The efficiency of the motor is 80%.
(a) What is the input power of the motor?
(b) Find the output power of the motor.
(c) What is the upward force needed to lift the bricks? (Take $g = 10$ N/kg.)
(d) How much energy is needed to lift the bricks?
(e) How long will it take the motor to lift the bricks?

15 A window-cleaner has a mass of 70 kg. He climbs up a ladder through a vertical distance of 6 m.
(a) How much work does he do?
(b) If he takes 5 s to climb the ladder, what is his power?

16 A light bulb is rated at 60 W. How much electrical energy does it convert in 20 minutes?

17 A crane lifts a crate of mass 500 kg by 20 m.
(a) What is the gravitational potential energy of the crate?
(b) What is the work done on the crate? (Take $g = 10$ N/kg.)

C7 18 Copy and complete the following:
(a) Force multipliers have a mechanical advantage _____ than 1.
(b) Distance multipliers have a velocity ratio _____ than 1.
(c) In general the VR of a pulley system equals the _____ of ropes supporting the lower moving block of pulleys.

19 Which of the following are force multipliers and which distance multipliers:
(a) fishing rod;   (b) crowbar;
(c) scissors;   (d) forearm;
(e) nutcracker;   (f) bicycle?

20 A block and tackle system is used to lift a load of 40 N. This is shown in the figure.

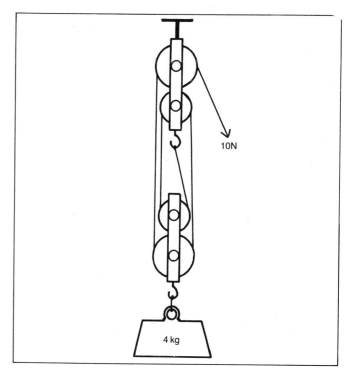

(a) What is the VR of this system?
(b) Why will the MA be less than 4?
(c) How would the efficiency change as the load increases?

21 An inclined plane is used to help load boxes on to a lorry. This is shown in the figure.

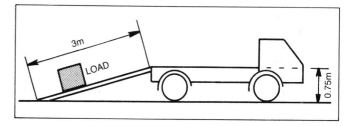

(a) What is the VR of the inclined plane?
(b) Is the plane a force multiplier or a distance multiplier?

8 Speed

The speed of an object is the distance it travels in a unit of time.

- The SI unit of speed is **metres per second (m/s)**. Other units in common use: kilometres per hour (km/h) and miles per hour (m.p.h.).

What is meant by an *average* speed?

A runner who completes a 400 metre lap of a running track in 50 seconds will vary her speed during the lap. We do not know her speed at any particular instant, but we can say that *on average* she runs 8 metres in each second. When distance s is measured in metres and time t in seconds, the average speed v is found in metres per second (m/s) using:

$$\text{average speed} = \frac{\text{distance moved}}{\text{time taken}} \qquad v = \frac{s}{t}$$

Example: A cyclist travels 54 km in 3.0 hours. Find his average speed in km/h and m/s.

In km/h: average $v = \dfrac{s}{t} = \dfrac{54\,\text{km}}{3.0\,\text{h}} = 18\,\text{km/h}$

In m/s: average $v = \dfrac{s}{t} = \dfrac{54\,000\,\text{m}}{3 \times 60 \times 60\,\text{s}} = 5\,\text{m/s}$

How do we measure speed?

Using a ticker-timer

A ticker-timer prints 50 dots every second on a strip of paper tape attached to a moving object. The small time interval Δt between dots is called a 'tick' and is 1/50 second or 0.02 s.

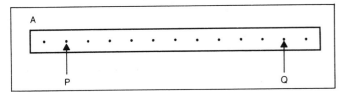

Figure 8.1 *Tape A:*

From dot P to dot Q there are ten spaces:

total time $t = \text{spaces} \times \Delta t = 10 \times 0.02\,\text{s} = 0.2\,\text{s}$

The distance gone from dot P to dot Q = 50 mm:

$\therefore$ average speed $v = \dfrac{s}{t} = \dfrac{50\,\text{mm}}{0.2\,\text{s}} = 250\,\text{mm/s or } 0.25\,\text{m/s}$

Measuring instantaneous speed

Suppose a moving object pulls tape through the ticker-timer. There is a small distance Δs between two dots. This is the distance moved by the object in the time interval Δt. The speed v at this moment is called the **instantaneous speed**: $v = \Delta s/\Delta t$:

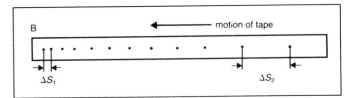

Figure 8.2 *Tape B:*

The instantaneous speed at the end of the tape is greater than it was at the beginning. The object has increased its speed or **accelerated**. This is shown by the fact that the dots *spread out*. So Δs_2 is greater than Δs_1.

How to make a tape chart

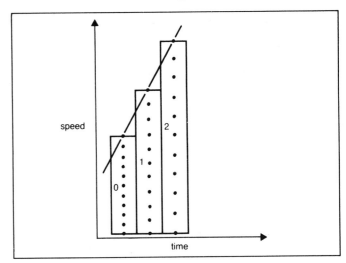

Figure 8.3:

This tape chart was made by attaching a tape to a trolley which was rolled down a sloping runway. The dots were marked out in groups of '*ten-ticks*'. The ten-ticks were numbered and cut up. They were then glued side by side to make the tape chart.

$$1 \text{ ten-tick} = 10 \times 0.02\,\text{s} = 0.2\,\text{s or } 1/5\,\text{s}$$

Each ten-tick gives a measure of the speed:

$$\text{speed} = \frac{\text{length of ten-tick strip}}{\text{time for ten-ticks}}$$

e.g. ten-tick number 2 is 40 mm long:

$\therefore$ speed at ten-tick no. 2 $= \dfrac{40\,\text{mm}}{0.2\,\text{s}} = 200\,\text{mm/s or } 0.2\,\text{m/s}$

The tape chart forms a **speed–time graph**. The graph shows increasing speed.

Speed–time graphs

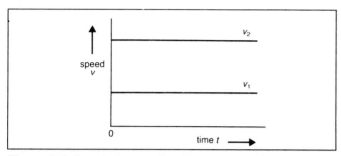

Figure 8.4 *Speed–time graphs of constant speed:*

Constant speed is shown by horizontal lines. An object has constant speed if it moves equal distances in equal time intervals. Speed v_2 is greater than speed v_1.

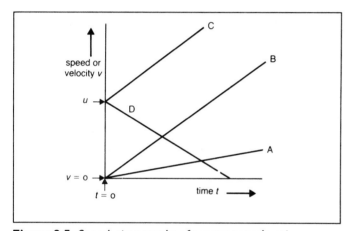

Figure 8.5 *Speed–time graphs of constant acceleration:*

A: Constant acceleration is shown by a constant upwards slope. The object starts from rest ($v = 0$).
B: This is faster constant acceleration from rest.
C: This is constant acceleration starting at a speed u.
D: This is constant deceleration from a speed u. The object comes to rest when the graph touches the time axis.

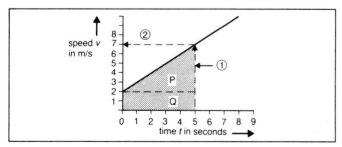

Figure 8.6 *Finding an instantaneous speed from a speed–time graph: at $t = 5$ seconds, speed $= 7$ m/s:*

The shaded area in figure 8.6 gives the distance moved:

$$\text{area} = \text{time} \times \text{speed} = \text{distance moved}$$

area P	$= \frac{1}{2} \times 5\,\text{s} \times 5\,\text{m/s}$	$= 12.5\,\text{m}$
area Q	$= 5\,\text{s} \times 2\,\text{m/s}$	$= 10\,\text{m}$
total distance moved $=$	$12.5 + 10$	$= 22.5\,\text{m}$

Distance–time graphs

Upwards slopes show increasing distance away. On distance–time graphs, constant slopes show constant speeds. The steeper the slope the faster is the motion.

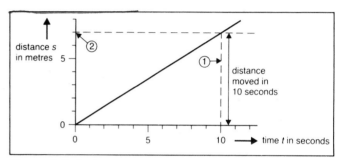

Figure 8.7 *Finding constant speed:*

The gradient of this graph gives the speed.

① Draw a vertical line at a suitable time (e.g. 10 s).
② Read the total distance moved at this time from the scale.

$$\text{constant speed } v = \frac{s}{t} = \frac{7\,\text{m}}{10\,\text{s}} = 0.7\,\text{m/s}$$

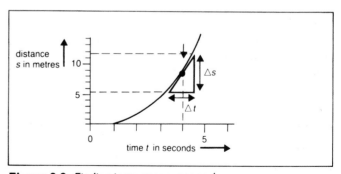

Figure 8.8 *Finding instantaneous speed:*

The gradient of the graph increases as the speed increases. Here a tangent is drawn at $t = 4$ seconds. This gives the instantaneous speed at 4 s. $\Delta s = 11.5\,\text{m} - 5.5\,\text{m} = 6\,\text{m}$. $\Delta t = 1$ second. So:

$$v = \text{gradient of tangent} = \frac{\Delta s}{\Delta t} = \frac{6\,\text{m}}{1\,\text{s}} = 6\,\text{m/s}$$

9 Velocity and acceleration

Which quantities are vectors?

Displacement is the distance moved *in a stated direction*.

Velocity is the speed *in a stated direction*.

Distance and speed are **scalars**.
Displacement and velocity are **vectors**.

Example: Think of a train travelling from London to York at a speed of 50 m/s. It can have the same *speed* as another train travelling from York to London, but it can never have the same *velocity*. The train from London has a velocity of 50 m/s *north*. However, the train from York has a velocity of 50 m/s *south*. Their different directions give them different velocities.

$$\frac{\text{average}}{\text{velocity}} = \frac{\text{displacement}}{\text{time taken}}$$

Acceleration

Acceleration is the rate of change of velocity with time. Acceleration is a vector.

$$\text{acceleration} = \frac{\text{change of velocity}}{\text{time taken}}$$

Δv = change of velocity
Δt = time interval
a = acceleration
$\left.\right\}$ $a = \dfrac{\Delta v}{\Delta t}$

An alternative formula is:

u = initial velocity
v = final velocity $\left.\right\}$ $\Delta v = v - u$
t = time taken to change
$\left.\right\}$ $a = \dfrac{v - u}{t}$

Example: A car has an initial velocity of 10 m/s. It accelerates to a final velocity of 40 m/s in 12 seconds. Find its acceleration.

$$\Delta v = (40 - 10)\,\text{m/s} = 30\,\text{m/s}$$
$$\Delta t = 12\,\text{s}$$
$$a = \frac{\Delta v}{\Delta t} = \frac{30\,\text{m/s}}{12\,\text{s}} = 2.5\,\text{m/s}^2$$

Units of acceleration

In the example above, every second the velocity increased by 2.5 m/s.

$\therefore$ The *rate of change* of velocity is 2.5 m/s *in each second*. The acceleration of the car is 2.5 metres per second *per second* or 2.5 m/s².

● The SI unit of acceleration is **metres per second squared (m/s²)**.

How do we find acceleration from a graph?

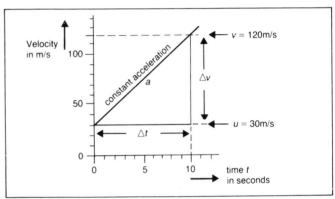

Figure 9.1 *Finding acceleration from a velocity–time graph:*

Acceleration is given by the *gradient* of a velocity–time graph. Here the initial velocity $u = 30$ m/s. The final velocity $v = 120$ m/s. $\Delta v = v - u = 120 - 30 = 90$ m/s. $\Delta t = 10$ s. So:

$$a = \frac{\Delta v}{\Delta t} = \frac{90\,\text{m/s}}{10\,\text{s}} = 9.0\,\text{m/s}^2$$

How do we use the equations of motion?

The equations of motion $\boxed{1}$ **to** $\boxed{4}$ apply to objects with constant acceleration. We can use them to work out their acceleration a, final velocity v, or displacement (distance moved) s.

$\boxed{1}$ $a = \dfrac{v - u}{t}$

$\boxed{2}$ $v = u + at$
$\boxed{3}$ $s = \text{average velocity} \times \text{time}$
$$s = \frac{(u + v)}{2}t$$
$\boxed{4}$ $s = ut + \tfrac{1}{2}at^2$

● The simple formula for distance is $s = vt$. This can be used only when v is constant.

Example: *Using equation* $\boxed{2}$. A cyclist starts from rest. He accelerates constantly at $1.5\,\text{m/s}^2$. Find his velocity after 8 seconds. Starting from rest means $u = 0\,\text{m/s}$. So:

$$\boxed{2} \qquad v = u + at$$
$$\therefore v = 0 + (1.5 \times 8) = 12\,\text{m/s}$$

Example: *Using equation* $\boxed{3}$. A car is travelling at $15\,\text{m/s}$. It accelerates to $27\,\text{m/s}$ in 8 seconds. How far does it travel?

$$\boxed{3} \qquad s = \frac{(u + v)}{2}\,t = \frac{(15 + 27)}{2} \times 8 = 168\ \text{metres}$$

Example: *Using equations* $\boxed{4}$ *and* $\boxed{2}$. An athlete runs up to a long jump. She accelerates at $1.6\,\text{m/s}^2$ for 6 seconds. Find the length of run-up she will need. What is her final velocity? Assuming she starts from rest, $u = 0\,\text{m/s}$. So:

$$\boxed{4} \qquad s = ut + \tfrac{1}{2}at^2$$
$$\therefore s = (0 \times 6) + (\tfrac{1}{2} \times 1.6 \times 36) = 14.4\ \text{metres}$$
$$\boxed{2} \qquad v = u + at = 0 + (1.6 \times 6) = 9.6\,\text{m/s}$$

How do objects fall?

A leaf falling from a tree will float to earth slowly because of the air resistance. In a vacuum a leaf would fall with the same acceleration as a stone. If they were released from the same height at the same time, both would hit the ground at the same time.

The acceleration with which things fall to Earth is called the **acceleration due to gravity *g*.**

$t = 0$ ◯	at rest	Near sea level *g* is about $10\,\text{m/s}^2$.
$t = 1\,\text{s}$	$v = 10\,\text{m/s}$	The velocity of a freely falling
$t = 2\,\text{s}$	$v = 20\,\text{m/s}$	object (of any size or weight) increases by $10\,\text{m/s}$ downwards every second. The velocity of an
$t = 3\,\text{s}$	$v = 30\,\text{m/s}$	object thrown upwards decreases by $10\,\text{m/s}$ every second until it reaches its maximum height.

Figure 9.2 *Acceleration due to gravity.*

Equations for falling objects

In the equations of motion, replace the symbol *a* with the symbol *g* or $10\,\text{m/s}^2$.

$\boxed{2}$ $v = u + gt$ or from rest $(u = 0)$: $v = 10t$

$\boxed{4}$ $s = ut + \tfrac{1}{2}gt^2$ or from rest: $s = 5t^2$

An object may be thrown *upwards* so that gravity causes a *deceleration*. Here use $g = -10\,\text{m/s}^2$.

Example: A parcel is dropped from an aircraft and hits the ground in 5 seconds. Assume its initial downwards velocity is zero. Calculate its velocity on impact and the height fallen.

$$u = 0\,\text{m/s} \quad \text{and} \quad g = 10\,\text{m/s}^2$$
$$\boxed{2} \quad v = gt \quad \text{or} \quad v = 10t = 10 \times 5 = 50\,\text{m/s}$$
$$\boxed{4} \quad s = \tfrac{1}{2}gt^2 \quad \text{or} \quad s = 5t^2 = 5 \times 25 = 125\,\text{m}$$

How do we measure *g* using a ticker-timer?

A mass of $200\,\text{g}$ falls and pulls a $2\,\text{m}$ length of tape through a timer. The tape chart is made using 'two-tick' lengths of tape as shown in figure 9.3. (These are used because anything more than two-ticks gives very long tapes.)

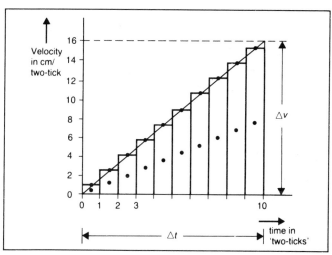

Figure 9.3 *Tape chart to find g:*

A straight-line graph is drawn through the tops of the tape strips to find the acceleration. The time interval for a two-tick strip is: $2 \times 0.02\,\text{s} = 0.04\,\text{s}$

$$\therefore \Delta t = 10 \text{ two-ticks} = 10 \times 0.04\,\text{s} = 0.4\,\text{s}$$

Each strip of tape is the distance fallen in two-ticks. So:

$$\Delta v = 16\,\text{cm/two-tick or } 16\,\text{cm in } 0.04\,\text{s}$$
$$\therefore \Delta v = \frac{16\,\text{cm}}{0.04\,\text{s}} = 400\,\text{cm/s}$$
$$\therefore a = \frac{\Delta v}{\Delta t} = \frac{400\,\text{cm/s}}{0.4\,\text{s}} = 1000\,\text{cm/s}^2 = 10\,\text{m/s}^2$$

In real experiments with ticker-tape the result is always less than $10\,\text{m/s}^2$. This is because of the drag on the tape as it goes through the timer. A freely falling object can be timed more accurately using multiflash or stroboscopic photography, or electronic timing.

Review questions: Chapters 8 and 9

C8

1. Copy and complete the following:
   (a) The *speed* of an object is the _____ it travels in a unit of time.
   (b) The speed at *a particular moment* is called the _____ speed.
   (c) A ticker-timer prints _____ _____ every second on a strip of paper.
   (d) The _____ of a distance–time graph equals the speed of the motion.

2. Which of the following are units of speed:
   (a) m/s;   (b) m.p.h.;   (c) km/s;   (d) J/s;
   (e) km/h;   (f) mm/s;   (g) Ns;   (h) cm/s?

3. Calculate the average speed of the following in (i) km/h; (ii) m/s:
   (a) a car which travels 108 km in 2 hours;
   (b) a bird which flies 1.8 km in 6 minutes;
   (c) a golf ball which goes 400 m in 20 s.

4. A ship steams at an average speed of 6 m/s. How far will it travel in:
   (a) 600 s;   (b) 5 hours?

5. The figure shows a strip of ticker-tape. The dots are printed every 0.02 seconds.

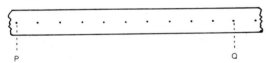

   (a) How many spaces are there from P to Q?
   (b) What is the total time from P to Q?
   (c) Measure the distance from P to Q.
   (d) Find the average speed in: (i) mm/s; (ii) m/s.

6. A trolley runs along the ground and pulls a strip of tape through a ticker-timer from left to right. The tape is shown in the figure.

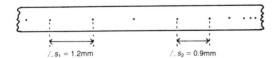

$\angle s_1 = 1.2\,\text{mm}$ $\angle s_2 = 0.9\,\text{mm}$

   (a) A dot occurs every 0.02 s. Find the instantaneous speeds, v_1 and v_2, for the distances Δs_1 and Δs_2 respectively.
   (b) Give a reason why the trolley slows down.

7. The figure shows a tape chart.

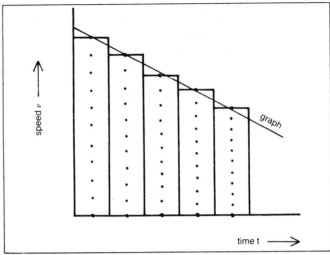

   (a) What is the name of the strips of tape?
   (b) What is the name of the instrument which makes the dots?
   (c) The time interval between dots is called a _____.
   (d) The time interval for ten spaces between dots is called a _____.
   (e) Describe the motion of the trolley.
   (f) What would the graph look like if the trolley had a constant speed?

8. The figure shows a tape chart of a trolley's motion. Each tick lasts 0.02 s. The strips have 10 ticks.

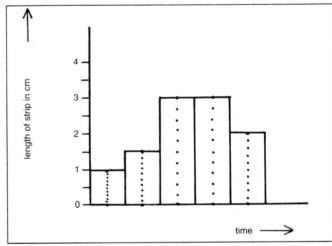

   What is:
   (a) the total distance travelled;
   (b) the total time taken;
   (c) the average speed of the trolley?

9. (a) Draw distance–time graphs to show:
       (i) a constant low speed; (ii) a constant high speed; (iii) a zero speed.
   (b) What would a horizontal line above the time axis mean?

10 Sketch speed–time graphs of the following.
Explain what the motion means in each case.
(a) a straight line through the origin with a gentle slope upwards;
(b) a straight line through the origin with a steep slope upwards;
(c) a horizontal line through the origin;
(d) a horizontal line close to the time axis;
(e) a horizontal line far from the time axis;
(f) a straight line starting half way up the speed axis and sloping downwards.

11 The figure shows a distance–time graph for part of a cyclist's journey.

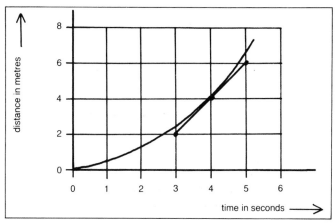

What is the instantaneous speed at 4 s?

C9 **12** The figure shows a speed–time graph for a car journey.

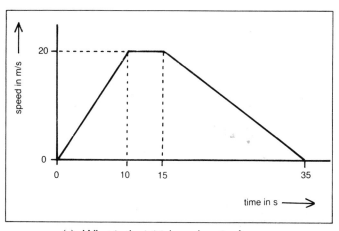

(a) What is the initial acceleration?
(b) How long does the car travel at a constant speed?
(c) What is the deceleration of the car?
(d) Find the total distance travelled.

13 Which of the following are vectors:
(a) distance;   (b) displacement;
(c) speed;   (d) velocity;   (e) acceleration?

14 What is the acceleration of:
(a) a car increasing its velocity steadily from 10 m/s to 15 m/s in 5 s?
(b) a train changing steadily from a velocity of 14 m/s to 30 m/s in 8 s?

15 Use the equations of motion to solve the following problems (assume all the accelerations are constant):
(a) An athlete starts from rest. She accelerates at 0.5 m/s² for 20 s. What is her final velocity?
(b) A speedboat accelerates from 6 m/s to 10 m/s in 3 s. How far will it have travelled?
(c) A train starts from rest. It accelerates at 0.6 m/s². After 10 s find:
  (i) its velocity;
  (ii) the distance travelled.

16 A labourer drops a spanner from the top of a tall building.
(a) What is the acceleration of the spanner?
(b) The spanner takes 2 s to reach the ground. What is the velocity on impact?
(c) How far has the spanner fallen?

17 A stone is thrown vertically upwards and lands on the ground 8 s later.
(a) How long does the stone take to reach its maximum height?
(b) What is the stone's velocity at this height?
(c) What is the velocity of the stone as it leaves the person's hand?
(d) What is the maximum height of the stone?

18 The figure shows a tape chart. This is the result of an experiment to find the value of g.

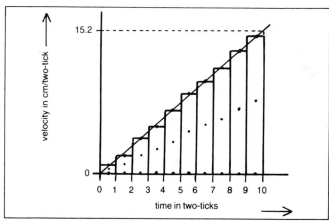

(a) Suggest a reason why 'two-tick' rather than 'ten-tick' lengths of tape are chosen.
(b) What is the time interval for a two-tick length?
(c) What is the time interval for 10 two-tick lengths?
(d) Use the graph to find g in cm/s².
(e) Convert your answer to m/s². Suggest a reason why the answer is less than 10 m/s²?

10

Newton's laws of motion

What does acceleration depend on?

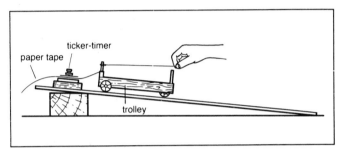

Figure 10.1 *Accelerating a trolley:*

This experiment investigates the relation between an *unbalanced* force applied to an object and the acceleration it causes.

The trolley is pulled down a runway. One, two and three stretched elastic cords are used in turn. The cords pull with equal forces F when stretched to the same length.

A gently sloping runway is used to compensate for friction. All the force F applied by each elastic cord can then go to producing acceleration. The slope is set so that when a trolley is given a push it runs down the slope at a constant velocity.

Results are shown below. The acceleration varies in direct proportion with the force:

$$a \propto F$$

A further experiment is done. Here trolleys are stacked to double and triple the mass. The results show that (for the same force) doubling the mass halves the acceleration, and so on.

Acceleration varies inversely with the mass:

$$a \propto 1/m$$

(a) With one cord: force = F; acceleration = $\dfrac{\Delta v}{\Delta t} = a$.

(b) With two cords: force = $2F$; acceleration = $\dfrac{2\Delta v}{\Delta t} = 2a$.

(c) With three cords: force = $3F$; acceleration = $\dfrac{3\Delta v}{\Delta t} = 3a$.

Newton's second law

The results of the above experiments can be combined to give an equation for **Newton's second law of motion:**

- acceleration = $\dfrac{\text{unbalanced force}}{\text{mass}}$ $a = \dfrac{F}{m}$

Rearranging this we get: $F = ma$

- The SI unit of force is the **newton (N)**.

- 1 newton is the unbalanced force which gives a mass of 1 kg an acceleration of 1 m/s^2.

As an example, take an unbalanced force of 4 N. This will give:
a 1 kg mass an acceleration of 4 m/s^2, or
a 4 kg mass an acceleration of 1 m/s^2.

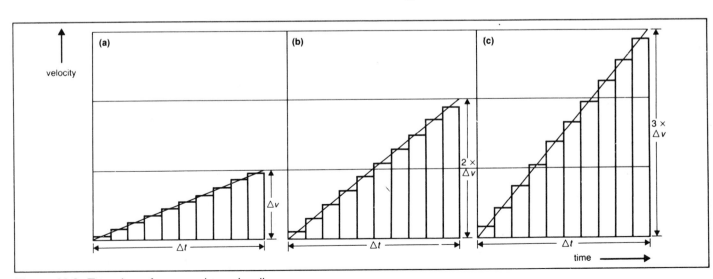

Figure 10.2 *Tape charts for an accelerated trolley:*

Newton's first law

- Stationary objects do not move on their own. They stay where they are unless an *unbalanced* force is applied to them.

- Similarly, moving objects do not change either their speed or their direction unless an *unbalanced* force is applied to them.

These statements express **Newton's first law.**

- Where there is *no* friction, no force is needed to keep an object moving and Newton's first law is seen to be true.

Example: A space ship in deep space does not use its engines to keep moving. Its engines are needed only to change its motion.

Figure 10.3 *An ice skater can glide over the surface of ice at almost constant speed in a straight line without any effort. This is because of low friction:*

- Where there *is* friction a force is needed to keep an object moving. It is harder here to see how the first law is obeyed.

Example: You must push your bicycle with a force to keep it moving along a level road. This is needed to overcome or balance out the frictional forces. Once friction is overcome, no extra or unbalanced force is needed to keep the bicycle going at a constant velocity.

How do objects fall against resistance?

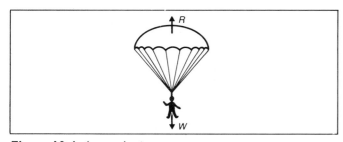

Figure 10.4 *A parachutist:*

- When $W = R$ the forces are balanced. Their resultant force is zero. Since a is zero, the parachutist falls at constant speed.
This motion is described by law 1.
- When $W > R$ there is an unbalanced force F:
$F = W - R$ downwards. By the second law:
$F = ma$. This gives $W - R = ma$.

The downwards acceleration is: $a = \dfrac{(W - R)}{m}$.

As the speed of the parachutist increases the air resistance also increases until it balances his weight. Then there is no more acceleration. The parachutist then has a *maximum* downwards velocity called a **terminal velocity.**

Newton's third law

When a force acts there are always *two* objects involved. We can call them object A and object B.

- If object A applies a force F to object B, then object B also applies a force $-F$ (of equal size but in the opposite direction) to object A.

- Forces always come in equal and opposite pairs which act on two separate objects.

- The two forces are equal in size and opposite in direction. They also act in the same straight line.

Examples of Newton's third law

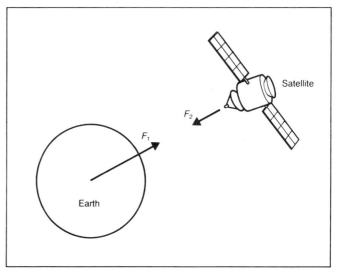

Figure 10.5 *Satellite in orbit round the Earth:*

The satellite pulls the Earth with a force which is equal in size, opposite in direction and in the same straight line as the force with which the Earth pulls the satellite.

A pair of magnets attract each other with equal and opposite forces. This is true even if the magnets are not of the same size or strength.

Kinetic energy and momentum

What is kinetic energy?

Motion energy is called kinetic energy.
Kinetic energy $E_k = \frac{1}{2}mv^2$.

● In SI units, mass m is in kg and speed v is in m/s. The kinetic energy is in **joules (J)**.

Example: A sprinter's mass is 60 kg. She runs at 10 m/s. What is her kinetic energy?

$$E_k = \frac{1}{2}mv^2 = 0.5 \times 60 \times 10^2 = 3000 \text{ joules}$$

How does work change kinetic energy?

Work must be done to get things moving. For example, a car engine does work converting the energy stored in petrol into kinetic energy (and some wasted heat energy).

$$\begin{array}{c} \text{work done on} \\ \text{a moving object} \end{array} = \begin{array}{c} \text{its change in} \\ \text{kinetic energy} \end{array}$$
$$Fs = \Delta(\tfrac{1}{2}mv^2)$$

● The kinetic energy depends upon the square of the speed. This means that to go *twice* as fast you need *four* times more energy.

Example: *Calculating the E_k gained from work done.* A free-wheeling motor cyclist has a mass (including her machine) of 100 kg. She is pushed from rest for 10 m. The push of 250 N acts against a frictional force of 70 N. Calculate her E_k and velocity when the push ends.

unbalanced force causing acceleration $=$ push $-$ frictional force
$$F = 250 \text{N} - 70 \text{N} = 180 \text{N}$$
the work done which causes acceleration $W = Fs = 180 \text{N} \times 10 \text{m} = 1800 \text{J}$

So the kinetic energy gained $E_k = 1800$ joules. Rearranging the formula for E_k gives:

$$v^2 = \frac{2E_k}{m} = \frac{12 \times 1800}{100} = 36$$
$$\therefore v = 6 \text{m/s}$$

How quickly can a car stop?

Work is done by the brakes of a car in slowing it down. It is equal to the amount of kinetic energy lost. This kinetic energy is converted by the brakes into heat energy.

● The *thinking distance* is the distance travelled at a constant speed before the brakes are applied. A driver who is alert takes at least 0.6 seconds to 'think' and press on the brake pedal. For example, at 15 m/s the distance he will travel is given by:

$$s = vt = 15 \text{m/s} \times 0.6 \text{s} = 9 \text{m}$$

● The *braking distance* is the distance travelled while the brakes are doing work turning the kinetic energy of the car into heat. If the braking force applied to the car is F, the work W done by the brakes is:

$$W = Fs = \Delta E_k = \Delta(\tfrac{1}{2}mv^2)$$

Compare the stopping distances of two cars X and Y. They are travelling at 15 m/s and 30 m/s respectively. The cars each have a mass of 960 kg. Their brakes can apply a braking force of 8 kN or 8000 N.

Using:	Fs	$= \Delta(\tfrac{1}{2}mv^2)$
For car X:	$8000s$	$= \tfrac{1}{2} \times 960 \times 225$
which gives:	s	$= 13.5 \text{m}$
For car Y:	$8000s$	$= \tfrac{1}{2} \times 960 \times 900$
which gives:	s	$= 54 \text{m}$

Car Y is travelling *twice* as fast as car X. But it has *four times* as much E_k and travels *four times* further during braking.

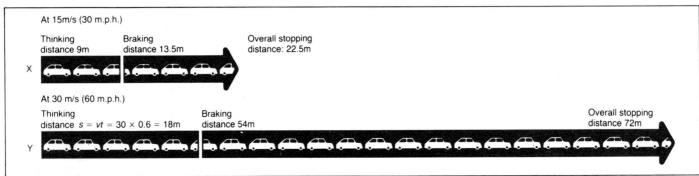

Figure 11.1 *Shortest stopping distances for an average family saloon car with good brakes on a dry road and an alert driver.*

What is momentum?

Momentum is mass × velocity. More mass more momentum, more velocity more momentum.

● Momentum $p = mv$

● The SI unit equals the unit of mass times that of velocity: $kg\,m/s$.
● Momentum is a *vector* quantity. It has the same direction as the velocity.

Example: A car has mass $1000\,kg$. It travels at a velocity of $20\,m/s$. The momentum is given by:

$$momentum = mv$$
$$= 1000\,kg \times 20\,m/s = 20\,000\,kg\,m/s$$

When is momentum conserved?

Momentum is a very useful concept in physics. It helps us to understand and calculate what happens in collisions and explosions. When two or more objects collide their total momentum is *constant* or *conserved* provided no external unbalanced force (e.g. friction) acts on them.

total momentum = total momentum
before collision   after collision

As momentum is a vector quantity, its direction must be taken into account. For example:

give motion → positive (+) momentum
and motion ← negative (−) momentum.

Example: *Collision.*

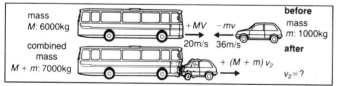

Figure 11.2:

momentum of
bus before $= +MV = +6000 \times 20 = +120\,000\,kg\,m/s$
collision

momentum of
car before $= -mv = -1000 \times 36 = -36\,000\,kg\,m/s$
collision

total
momentum $= +120\,000 - 36\,000 = +84\,000\,kg\,m/s$
before collision

After the collision the bus and car move together with a mass of $7000\,kg$. The new velocity is v_2. Since it is conserved the total momentum is $+84\,000\,kg\,m/s$.

Total momentum
after collision $= (M + m)v_2 = 7000v_2 = +84\,000\,kg\,m/s$

so $v_2 = \dfrac{+84\,000\,kg\,m/s}{7000\,kg} = +12\,m/s$

How do we calculate the effect of an explosion?

Momentum is conserved in explosions. Examples are those which occur in a gun or a rocket.

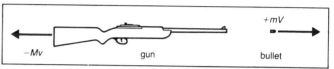

Figure 11.3 *Firing a gun:*

Before a gun is fired both the gun and bullet have zero momentum. After firing, the gun and bullet have an equal and opposite momentum. Their combined momentum therefore remains zero. The bullet has small mass but high velocity. The gun has large mass and small recoil velocity.

$$-Mv + mV = 0$$

Example: A girl has mass $50\,kg$. She leaps out of a small boat of mass $250\,kg$ on to a jetty. The girl's horizontal velocity is $2\,m/s$. With what velocity will the boat move away from the jetty?

momentum + momentum = 0 before and after
of girl   of boat   the leap

$(50\,kg \times 2\,m/s) + (250\,kg \times v) = 0$

$= -\dfrac{(50 \times 2)}{250}\,m/s = -0.4\,m/s$ (opposite direction)

How do impulses change momentum?

The momentum of an object is constant only when no resultant force acts on it.

An unbalanced force F acting on an object for a time interval Δt causes a *change of momentum*. The force equals the rate of change of momentum. This is another form of Newton's second law of motion or $F = ma$.

$$force = \frac{change\ of\ momentum}{time} \qquad F = \frac{\Delta(mv)}{\Delta t}$$

Multiplying both sides by Δt we get:

$$F\,\Delta t = \Delta(mv)$$

Force × time (or $F\Delta t$) is called an impulse.

● The SI unit of impulse is **newton seconds (Ns)**.

Eggs are packed in soft, shock-absorbing boxes so that when they suddenly stop or start moving they do not get cracked. In a hard box a large force acting for a short time cracks the egg. In a soft box a weaker force lasting for a longer time does not crack the egg. The same impulse acts in each case.

12 Transport

What are gears used for?

In most forms of wheel-driven transport the engine or power source is linked to the wheels by a system of gears. The gears allow the speed of rotation of the wheels to be changed in relation to the engine speed.

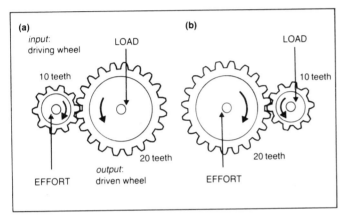

Figure 12.1 *Gears:*

The larger wheel with more teeth always turns more slowly. Gears in contact turn opposite ways round.

$$\frac{\text{rotation speed, larger wheel}}{\text{rotation speed, smaller wheel}} = \frac{\text{teeth on smaller wheel}}{\text{teeth on larger wheel}}$$

$$VR = \frac{\text{number of teeth on the driven wheel}}{\text{number of teeth on the driving wheel}}$$

(a) Slowing down rotation. VR of axles = 2 and MA > 1 (but < 2). The 10-teeth input axle turns quickly. The 20-teeth output axle turns slowly. This produces a 'low' gear. The wheels turn slowly and the gears multiply the effort of the engine or cyclist. A 'low' gear is required when setting off or when climbing a hill.

(b) Speeding up rotation. VR of axles = $\frac{1}{2}$ and MA < 1. The 20-teeth input axle turns slowly. The 10-teeth output axle turns quickly. This produces a 'high' gear. The speed of rotation is multiplied by 2. So this gear is a distance or speed multiplier. It is used to make a vehicle travel faster.

Example: In its top gear a car can go 3.5 times faster than in its first gear for the same engine speed. Suppose the speed of the car is 5.0 m/s in first gear when the engine is turning at 2000 r.p.m. How fast will it travel in top gear for the same engine speed?

The higher or 'top' gear multiplies the road speed (and distance travelled) by 3.5.

$$\therefore \text{speed in top gear} = 3.5 \times 5.0 \,\text{m/s} = 17.5 \,\text{m/s}$$

Why use gears on a bicycle?

We want a bicycle to move at a reasonable speed. So the wheels must go round faster than the pedals. A bicycle gear system is designed as a speed multiplier.

Example: If the chain wheel has 48 teeth and the free-wheel (fitted to the rear wheel) has 16 teeth, the VR is 1/3. If a cyclist turns the chain wheel round once every second, the rear wheel will turn three times a second.

If the rear wheel has a diameter of 0.7 m, for each revolution the bicycle moves $\pi \times 0.7$ m (the circumference of the wheel) = 2.2 m.

$$\therefore \text{the speed is } 3 \times 2.2 \,\text{m/s} = 6.6 \,\text{m/s}$$

Why do we use a lever for bicycle brakes?

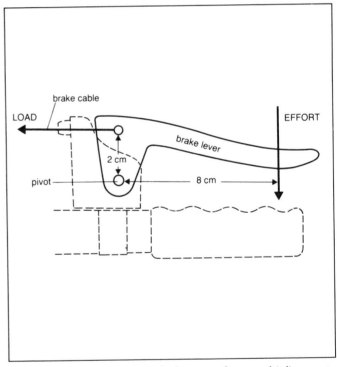

Figure 12.2 *The bicycle brake lever – a force multiplier:*

The hand-brake lever multiplies the effort. This is similar to a crowbar. The length of the lever allows the effort to move about four times further than the cable which pulls the brakes (the load). The lever moves 8 cm while the brake cable moves 2 cm. This gives a VR of: 8 cm/2 cm = 4. The MA is about 4. If the effort applied by hand is 30 newtons, the tension in the brake cable will be: $4 \times 30 = 120$ newtons.

What do performance figures for cars tell us?

Example: A manufacturer may claim that a modern car can accelerate from 0 to 30 m/s ($\approx$ 60 m.p.h.) in 10 s. If the car has a mass of 800 kg, find:
(a) the acceleration of the car;
(b) the force with which the engine accelerates the car;
(c) the power of the car.

(a) $a = \dfrac{\Delta v}{\Delta t} = \dfrac{30 \text{ m/s}}{10 \text{ s}} = 3.0 \text{ m/s}^2$.

(b) $F = ma = 800 \text{ kg} \times 3.0 \text{ m/s}^2 = 2400$ newtons.
The total force will be much greater because it has to overcome all the friction as well.

(c) Work done = gain of kinetic energy: ΔE_k.

$\Delta E_k = \Delta(\tfrac{1}{2}mv^2) = \tfrac{1}{2} \times 800 \times (30)^2 = 360\,000$ joules

The power output of the car equals the kinetic energy gained per second:

$P = \dfrac{\Delta E_k}{\Delta t} = \dfrac{360\,000 \text{ J}}{10 \text{ s}} = 36\,000 \text{ J/s or } 36 \text{ kW}$.

How can we compare different forms of transport?

The simplest way is to compare how far different forms of transport go for a unit of energy. Figure 12.3 compares distances in metres per megajoule of energy. On this chart the car appears to be by far the most economical. However, it is misleading because we should take into account the number of passengers carried. Figure 12.4 is obtained by multiplying the distances by the number of passengers. This shows that a well-loaded train is the most economical.

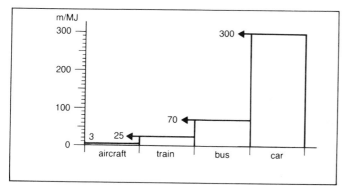

Figure 12.3 *Energy consumption for passenger transport (in metres per megajoule):*

Divide by 10 to convert to miles/gallon.
Divide by 30 to convert to kilometres/litre.

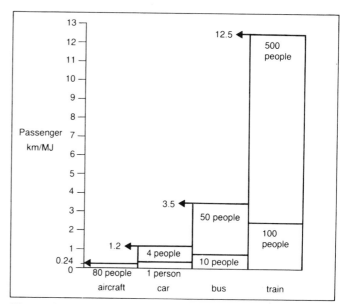

Figure 12.4 *Passenger distance travelled (kilometres) per megajoule:*

Multiply by 100 to convert to passenger miles/gallon.
Multiply by 100/3 to convert to passenger kilometres/litre.

Comparing forms of transport.

Vehicle	Advantages	Disadvantages
Bicycle	No pollution, no fossil fuel, good exercise, cheap travel	Range limited by time and fitness, some accident risk, unpleasant in bad weather
Car	High speed, convenience, comfort	Air and noise pollution, greedy for fuel, some accident risk
Bus	For 20+ people more economical than car, safe travel, comfort	Limited to bus routes, uses fossil fuel, air and noise pollution
Train	For 50+ people most economical, very safe, high speed, comfort	Limited number of stations, diesel trains use fossil fuel and pollute
Electric train	No pollution, near 100% energy conversion to E_k	
Aircraft	Very high speed, comfort, very safe	Limited to airports, very greedy for fuel, use fossil fuel, air and noise pollution

Review questions: Chapters 10, 11 and 12

C10 1 Copy and complete the following:
(a) Acceleration varies in _____ proportion to the *force*.
(b) Acceleration varies in _____ proportion to the *mass*.
(c) 1 newton is the _____ force which gives a mass of 1 kg an acceleration of _____.
(d) Stationary objects move if an _____ force acts on them. Objects change their speed or direction if an _____ force acts on them.

2 A cyclist pushes his bicycle with a force of 4 N and they move at a constant speed.
(a) How big is the frictional force on the bicycle?
(b) If he pushed with a force of 8 N what would happen to his speed?

3 A girl of mass 60 kg jumps out of an aeroplane. Before she opens her parachute her acceleration is 6.0 m/s².
(a) What is her weight?
(b) What is the unbalanced force on her?
(c) What is the frictional drag which is reducing her acceleration?
(d) When the acceleration is zero, what is her final velocity called?
(Assume $g = 10$ m/s².)

4 (a) Copy and complete the following sentence: Forces always come in equal and _____ pairs which act on two separate objects.
Peter pushes against a tree:
(b) Suppose he pushes with a force of 100 N. Give the name and size of the force which the tree exerts on Peter.
(c) What would happen if Peter's force were bigger than the tree's force?

C11 5 Calculate the kinetic energy of:
(a) a girl of mass 50 kg running at 10 m/s;
(b) a bumble bee of mass 4 g flying at 1 m/s;
(c) a snail of mass 0.01 kg moving at 1 mm/s.

6 By how many times must a boy's kinetic energy be increased if he wants to run:
(a) twice as fast; (b) three times as fast?

7 A fireman has a mass of 70 kg. He slides down a pole of height 10 m. The frictional force is 14 N.
(a) What is his weight?
(b) What is the resultant downwards force?
(c) Find the work done which causes acceleration.
(d) Calculate his velocity just before he lands at the bottom of the pole.

8 Highway Code tables give information about stopping distances for cars. The data are also given as 'thinking distance' and 'braking distance'.
(a) How do these two distances combine to give a stopping distance?
(b) What is meant by 'thinking distance'?
(c) What is meant by 'braking distance'?

9 A driver takes 0.8 s to react to the brake lights of the car in front and press on her brake pedal. If she is travelling at 20 m/s, what is her 'thinking distance'?

10 The table shows stopping distances for cars.

Speed in m/s	Distances in m		
	Thinking	Braking	Stopping
10	6	6	12
20	12	24	36
30	18	54	72
40			

(a) A car of mass 1000 kg is 'speeding' by travelling at a speed of 40 m/s. If the brakes apply a braking force of 8000 N, find the stopping distance of the car.
(b) Study the data in the table. Find the 'thinking distance' for a speed of 40 m/s.
(c) Complete the table for a speed of 40 m/s.

11 Calculate the momentum of:
(a) a tennis ball of mass 0.1 kg moving at a speed of 4 m/s;
(b) a javelin of mass 0.6 kg moving at 10 m/s;
(c) an elephant of mass 1400 kg walking at 2 m/s.

12 The figure shows two cars in a collision.

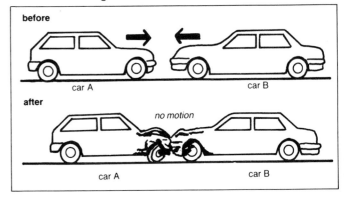

Car A has a mass $m_A = 800\,kg$ and a velocity $v_A = 30\,m/s$. Car B has a mass $m_B = 1000\,kg$ and a velocity v_B. For this collision:
(a) What is the total momentum afterwards?
(b) What is the total momentum before?
(c) Write an equation for the momentum.
(d) Use the equation to find the velocity v_B.

13 A man fires a rifle at a target. The combined mass of the man and the rifle, $M = 70\,kg$. The bullet has a mass, $m = 30\,g$. Its velocity as it leaves the rifle, $v = 700\,m/s$. The recoil velocity of the man plus the rifle $= V$.
(a) Write a momentum equation for this situation.
(b) Calculate the recoil velocity of the man plus the rifle.

14 Megan catches a cricket ball. The ball is travelling at a speed of $20\,m/s$. It has a mass of $0.5\,kg$. Find:
(a) the ball's momentum just before it stops moving;
(b) the ball's change of momentum;
(c) the average force exerted on Megan's hands by the ball if she stops it in $0.1\,s$.
(d) Megan pulls her hands in towards her body as she catches the ball. Describe:
 (i) how the time to stop the ball changes;
 (ii) how this changes the force exerted on her hands by the ball.

15 A stone of mass $10\,kg$ falls from a cliff. It has a speed of $54\,m/s$ just before it stops. It lands in sand without bouncing and stops in 0.5 seconds. What is the average force exerted on the stone by the sand?

C12 16 Copy and complete the following:

(a) In a set of gears, the _____ wheel with more teeth turns more slowly.

(b) Gears in contact turn _____ ways.

17 The figure shows a set of gears. Assume the large wheel is the driving wheel.

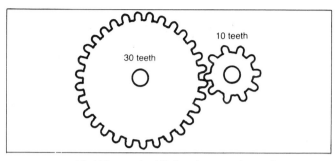

30 teeth   10 teeth

(a) What is the VR for this set of gears?
(b) Will the MA be greater or less than 1?
(c) Is this called a low or high gear?

18 (a) The chain wheel of a bicycle has 48 teeth. The freewheel (on the rear wheel) has 16 teeth. What is the VR?
(b) If a cyclist turns the chain wheel round once every second, how many times does the rear wheel turn every second?
(c) The diameter of the rear wheel is $0.8\,m$. How far does the bicycle move for each revolution?
(d) What is the speed in this gear?

19 (a) Is a bicycle brake lever a force multiplier or a distance multiplier?
(b) If the pivot is $2\,cm$ from the brake cable and $10\,cm$ from the position of the hand, what is the VR of the system?
(c) Will the hand (effort) move more or less than the cable (load)?

20 Use figures 12.3 and 12.4 on page 31 to help you answer the following:
(a) How many metres does a car travel per megajoule of energy?
(b) Convert this into miles/gallon.
(c) Repeat parts (a) and (b) for a train.
(d) The car and the train both have four passengers. How many passenger miles will be travelled per gallon for each? Which of these is the most economical use of energy?
(e) How many passengers must the train carry to have the same energy consumption per passenger as the car with four passengers?

Flight and orbits

Flight

Figure 13.1 *The forces on a thrown object:*

Only two forces act on an object thrown through the air:
  the weight *W* downwards;
  the air resistance *R* backwards.
  The path through the air is a parabola.

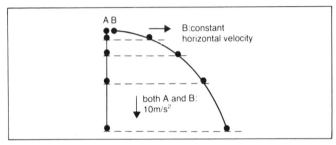

Figure 13.2 *Horizontal and vertical motions are independent:*

A ball A is dropped vertically. At the same moment another ball B is thrown sideways. They both have the same vertical motion. Both have a vertical acceleration of $10\,\text{m/s}^2$. B also has a constant horizontal velocity. The horizontal motion of B does not affect its vertical motion.

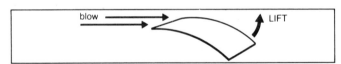

Figure 13.3 *Demonstrating the Bernoulli effect:*

When air is blown over the top of a sheet of paper, the paper rises in the air stream. This happens because the pressure *falls* above the paper where the air is moving faster.

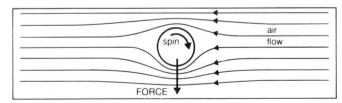

Figure 13.4 *The flight of a spinning ball:*

Below the ball the air flows faster as it is dragged round with the spin of the ball. The faster air flow reduces the pressure below the ball. This causes the flight of the spinning ball to curve downwards towards the lower pressure.

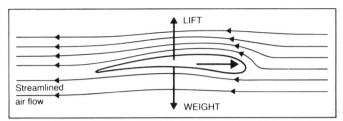

Figure 13.5 *Lift from an aerofoil:*

The lift on the wing of an aircraft is due to the Bernoulli effect. In this case there is lower pressure above the wing. This is due to the faster air flow there. The lift must balance the weight of the aircraft to keep it up in the air.

The turbo-jet or gas–turbine engine

- A jet engine needs a supply of air and so it cannot work in space.
- A gas–turbine engine works as follows:
  1. Air is sucked in.
  2. The air is compressed by fans.
  3. Fuel (kerosene) is injected.
  4. The fuel–air mixture burns explosively.
  5. The jet of hot gas turns the turbine fan.
  6. The turbine fan turns the compressor fan.
  7. The high-speed exhaust jet of gases thrusts the engine and aircraft forwards. This illustrates Newton's third law of motion.

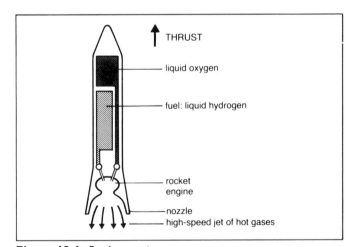

Figure 13.6 *Rocket engine:*

- Rocket engines carry their own oxygen. So they do not need air and can work in space.
- The gain of momentum of the rocket is equal and opposite to the momentum of the ejected hot gases. This is Newton's third law of motion.
- Liquid fuel rocket engines can be started and stopped. Solid fuel rockets cannot be shut down once they are ignited.

In the rocket shown in figure 13.6 there is liquid oxygen in the top tank. It is kept at −183°C. (There are 600 tonnes in the space shuttle's main tank.)

The fuel is liquid hydrogen. It is kept at −253°C. (There are 100 tonnes in the shuttle's main tank.) Some rockets use kerosene as a fuel instead.

The pumps inject fuel and oxygen into the engine. Here the fuel burns explosively with the oxygen. This produces a high-speed jet of gases.

What makes an object move in a circle?

When no unbalanced force acts on an object it will continue moving in a straight line at a constant speed. This is as described by Newton's first law. Motion in a circle needs an unbalanced force acting towards the *centre* of the circle. This keeps pulling the object into the curved path away from its natural straight line motion.

An *inward-acting* force is called centripetal.

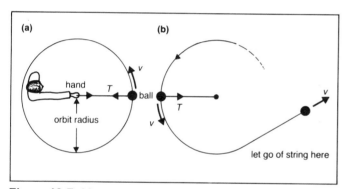

Figure 13.7 *Motion in a circle:*

(a) A ball is whirled round in a circle. It needs an inward or centripetal force to stay in orbit. The tension T in the string pulls inwards on the ball. An equal and opposite force pulls outwards on the hand.

(b) What happens when you let go? The ball flies off at a tangent to the circle. It now moves in a straight line at constant speed v. This is because no force acts on it.

Examples

● The Moon and satellites are held in orbit around the Earth by the gravitational (inwards-acting) pull of the Earth.
● An electron is held in orbit around an atomic nucleus by the attraction between unlike electric charges.
● The inwards push of the rotating drum wall makes clothes in a spin-dryer go round in a circle. (Water escaping through holes in the drum flies off at a tangent to the circle.)
● When you are whirled round on a roundabout your seat pushes you from behind towards the centre of the roundabout.

How can we calculate an orbit?

The gravitational force acting on a satellite must exactly provide the centripetal force needed to keep it in orbit. (See satellite B in figure 13.8 below.) The centripetal force needed = mv^2/r where m is the mass of the satellite, v is its speed and r the radius of its orbit.

So a larger force is needed if:
● the speed of the satellite increases; or
● the radius of its orbit decreases.
Gravity is stronger nearer the Earth. So satellites in low orbits must travel faster than those in higher orbits.

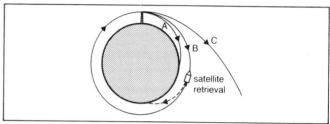

Figure 13.8 *Satellites:*

Imagine that three satellites could be launched horizontally from a tower reaching high above the Earth's atmosphere.

Satellite A is too slow to stay in orbit. It falls to the ground.

Satellite B is launched at orbital speed of about 8 km/s. Although 'falling' all the time, it never comes any closer to the ground. To retrieve the satellite the rocket engines are retro-fired. This reduces the speed. The satellite then falls back to Earth.

Satellite C is going too fast to stay in this orbit. The gravitational force is less than the centripetal force needed. So it goes off into space.

Geosynchronous satellites

Some satellites are placed in a high orbit (about 35 000 km above the equator). There they travel round the orbit once every 24 hours. So they appear to 'hover' above the same place on Earth. These Earth-synchronised satellites are used for permanent communication links.

When do we feel weightless?

Out in deep space, far from the gravitational attraction of any stars or planets, an object would be truly *weightless*. You feel 'weightless' while you are in the air after jumping off a springboard or a trampoline. This is because there is no contact force between you and the ground. However, your weight pulls you back down to the ground. So you are not truly weightless. An astronaut in orbit around the Earth *feels* 'weightless'. This is because both he and the spaceship are in **free fall** towards the Earth. There is no contact force between him and the spaceship. But the astronaut does have weight which provides the centripetal force needed to keep him in orbit.

The solar system

The Sun

Our Sun is one of 10^{11} stars in our Galaxy. We see our Galaxy as the Milky Way.

Distances

The speed of light $c = 3 \times 10^8$ m/s. The Earth is 150 million km from the Sun. So the time it takes for light to reach us from the Sun is given by:

$$t = \frac{s}{c} = \frac{150 \times 10^9 \, \text{m}}{3 \times 10^8 \, \text{m/s}} = 500 \, \text{s} = 8.3 \, \text{min}$$

A light-year is the distance light travels in 1 Earth year.

This can be calculated using $s = vt$:

$$s = ct = 3 \times 10^8 \times 60 \times 60 \times 24 \times 365 \, \text{m} = 9.5 \times 10^{15} \, \text{m}$$

This is 9.5×10^{12} km or 9.5 million million kilometres. It is 4.2 light-years to Proxima, the nearest star. The Milky Way is about 100 000 light-years across!

The power source of the Sun

The Sun is powered by **nuclear fusion.** The core of the Sun is at a temperature of 14×10^6°C. At this temperature hydrogen nuclei can fuse together to form helium nuclei. This fusion process releases the huge amounts of energy which keep the Sun shining.

Radiation from the Sun

The Sun emits radiation across a wide range of the electromagnetic spectrum. As well as the light we see and the infrared (heat radiation) we feel, the Sun also emits radio waves and ultraviolet radiation. A 'solar wind' also carries millions of small particles out into space. This wind 'blows' the tails of comets to point away from the Sun. It also bombards Earth's outer atmosphere. This creates the ionosphere where the atoms are all ionised.

The planets

Mercury: the inner planet (orbits in 88 days)

Mercury has no atmosphere. Its cratered surface reaches +350°C (hot enough to melt lead) in the daytime and falls to −170°C at night. Being close to the Sun, Mercury is strongly attracted by the Sun's gravitation. To stay in orbit and not be pulled into the Sun, Mercury travels at 48 km/s (the fastest of the planets).

Venus: the cloudy planet (orbits in 225 days)

Venus appears like a brilliant star. However, its light, reflected by its white clouds, all comes from the Sun. Venus has phases like the Moon. The atmosphere of Venus is made mostly of carbon dioxide. This holds in the heat making its surface temperature rise to 480°C. This is even hotter than Mercury's surface.

Mars: the red planet (orbits in 687 days)

Mars has an oval orbit with no tilt. This means winter occurs all over Mars at the same time when it is furthest from the Sun. Mars has two tiny moons. They are called Phobos and Deimos.

Jupiter: the largest planet

Jupiter consists mostly of the lightest element hydrogen, although it is easily the most massive of all the planets. Its outer atmosphere contains compounds of hydrogen such as water and ammonia. Jupiter has fifteen moons. It also has a very thin ring system.

Saturn: The ringed planet

The structure of Saturn is very similar to that of Jupiter. Saturn has many rings. These are composed of millions of rocks of varying size. It has seventeen moons.

The outer planets: Uranus, Neptune and Pluto

Uranus and Neptune are probably similar in composition to Saturn. They are far from the Sun, so they are very cold and dark. Pluto is probably an ice-covered rocky planet. Pluto's orbit is tilted and eccentric. This has brought the planet closer to the Sun than Neptune until 2009. Uranus has a few narrow rings spaced far apart.

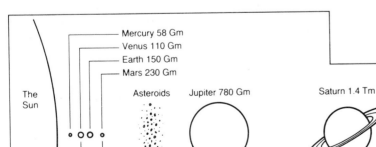

Mercury 58 Gm
Venus 110 Gm
Earth 150 Gm
Mars 230 Gm

The Sun
Asteroids
Jupiter 780 Gm
Saturn 1.4 Tm
Uranus 2.9 Tm

[110] [0.4] [1] [0.5] [1] [11] [9] [4]

The Earth: a watery world (orbits in 365 days)

The age of the Earth is about 4700 million years. Its diameter is 13×10^6 m. Its mass is 6.0×10^{24} kg. The Earth rotates in 24 hours. Its orbit speed is 30 km/s. The radius (distance from the Sun) of the Earth's orbit is 1.5×10^{11} m. Its mean density is 5.5 g/cm^3.

The surface of the Earth is 70% water and 30% land. Its atmosphere is composed of: 78% nitrogen, 21% oxygen, 0.9% inert gases, 0.07% water vapour and 0.03% CO_2.

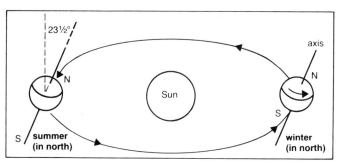

Figure 14.2 *The seasons:*

There is a $23\frac{1}{2}°$ tilt in the Earth's axis. This gives us the seasons as we orbit the Sun. The northern hemisphere has its summer when it is tilted towards the Sun (on the left of the figure).

The *Earth's crust* is only 5 km thick at the bottom of the oceans. It is 30 km thick under the continents. The crust is like the skin on an apple.

The *core* is probably liquid nickel–iron. However the centre may be solid due to the very high pressure. Here the temperature may reach 2200 °C. Rotation of the liquid iron core sets up large electric currents which generate the Earth's magnetic field.

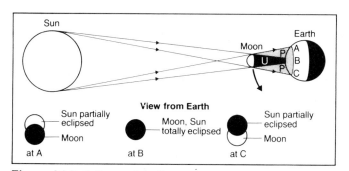

Figure 14.3 *Eclipses of the Sun:*

As the Moon orbits the Earth it occasionally passes between the Sun and the Earth. This casts a shadow on the Earth in the daytime.

In the **umbra** (U) of the shadow there is total darkness (point B). Here a *total eclipse* is seen. In the **penumbra** (P) of the shadow only part of the Sun is eclipsed (points A and C).

The Moon: a silent world

The Moon's age is probably the same as the Earth's. Its diameter is 3.5×10^6 m. The mass is 7.4×10^{22} kg. Gravity at the Moon's surface is only $0.165g$ or 1.6 N/kg. The radius of orbit or distance from the Earth varies between 350 and 400 Mm (million metres). The Moon orbits the Earth once every 29.5 days and also rotates on its own axis once every 29.5 days. So the same side of the Moon always faces the Earth.

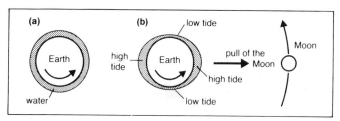

Figure 14.4 *The tides:*

The gravitational attraction of the Moon and the Sun both affect the Earth and make the tides. Being near the Earth, the Moon has a greater effect.

(a) The layer of water around the Earth would be at a constant height if it were not attracted by the Sun or Moon.

(b) The water on the side of the Earth nearer the Moon is pulled up into a high tide by the strong attraction of the Moon. On the opposite side of the Earth another high tide happens. This is because this water is furthest from the Moon. So it is not so strongly pulled in by the combined gravity of the Earth and the Moon.

When the Moon and the Sun are nearly in line with the Earth they pull together. This makes the largest tides called **spring tides.**

Neap tides are smaller tides. They occur when the Moon and the Sun are pulling at right angles and so work against each other.

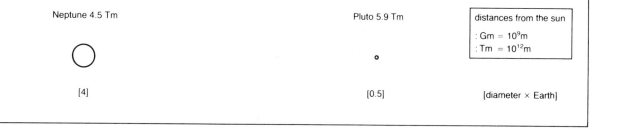

Review questions: Chapters 13 and 14

1 Copy and complete the following:
A javelin is thrown through the air:
(a) The path it follows is called a _____.
(b) The horizontal and vertical motions are _____.
(c) The force which acts *downwards* is _____.
(d) The force which acts *backwards* is _____.

2 A ball is thrown horizontally from an upstairs window of a house. Ignoring air resistance, it takes 2 s to reach the ground.
(a) Describe the shape of the path of the ball as it moves through the air.
(b) If the ball was then dropped vertically from the window, how long would it take to reach the ground?

3 Copy and complete the following:
Pressure *falls* when streamlined air or fluid moves _____. This is called the _____ effect. A cricketer puts spin on the ball to make its flight _____. When the ball spins the pressure difference causes a _____ on the ball.

4 (a) Sketch a diagram of an aeroplane flying in level flight. Show where the air speed is greatest. Show where the pressure is greatest.
(b) Which force balances the lift F?
The aeroplane has a mass $m = 40\,000$ kg. Its total wing area $A = 100\,\text{m}^2$. Δp is the difference in pressure between the upper and lower surfaces of the wings. So $\Delta p = F/A$.
(c) Calculate Δp. (Assume $g = 10$ N/kg.)

5 (a) Why can gas–turbine engines not be used in space?
(b) Describe the process which makes the compressor fans turn in a gas–turbine engine.
(c) What is the mixture which burns explosively?
(d) Name the equal and opposite forces produced in a gas–turbine engine.

6 (a) Rocket engine fuel can be liquid or solid. Which type can be started and stopped?
(b) Why can rockets be used in space whereas gas–turbine engines cannot?
(c) Why might ice form on the outside of the rocket?

7 Copy and complete the following:
(a) An object moving in a straight line with a constant speed will continue if _____ unbalanced force acts on it.
(b) To move in a circle the object needs an _____ force which must act towards the _____ of the circle.
(c) A force which makes an object move in a circle is called _____.

8 What provides the inward-acting force in each of the following:
(a) the Moon orbiting the Earth;
(b) an electron orbiting a nucleus;
(c) a car turning a corner;
(d) a mass whirled round on the end of a rope?

9 Calculate the centripetal force in the following:
(a) a 1 kg stone tied on the end of a rope and moving at 2 m/s in a horizontal circle of radius 2 m;
(b) a 60 kg skater moving in a circle of radius 10 m at a speed of 3 m/s.

10 Describe what happens when the centripetal forces in question 9 are removed.

11 The figure shows a satellite orbiting the Earth. When the satellite is in a stable orbit, the gravitational force provides the centripetal force. Describe what would happen:

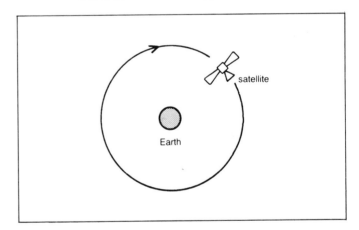

(a) if the gravitational force were:
(i) bigger; (ii) smaller;
(b) if the satellite's speed became smaller?

12 Why does an astronaut feel weightless when orbiting the Earth in a spaceship? If he drops a cup why does it seem to 'float'?

13 Jupiter is 780 million km from the Sun. The speed of light is 3×10^8 m/s. How long will light take to reach Jupiter?

14 Which planet(s):
  (a) is closest to the Sun;
  (b) is largest in diameter;
  (c) is called the red planet;
  (d) are surrounded by rings;
  (e) has a tilted, eccentric orbit;
  (f) moves with the greatest speed?

15 (a) What is meant by a light-year?
  (b) Light travels at 3×10^8 m/s. How far is a light-year in kilometres?

16 The bar chart shows the diameter of some of the planets.

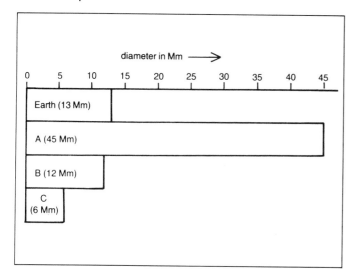

The planets are Venus, Earth, Neptune and Pluto. The top bar represents Earth's diameter (13 million metres). Which planet is represented by bar:
(a) A; (b) B; (c) C?

17 (a) How is the Sun powered?
  (b) Which of the following is the most likely temperature of the Sun's core:
    (i) 100°C; (ii) −180°C; (iii) 14×10^6°C; (iv) 18×10^{20}°C; (v) 0°C?
  (c) Name three types of radiation which are emitted by the Sun.

18 (a) Which two planets are nearest to the Earth?
  (b) Which gas makes up over three-quarters of the Earth's atmosphere?
  (c) What percentage of the Earth is land?
  (d) What is the core made of?
  (e) How is the Earth's magnetic field generated?

19 The figure shows the Earth orbiting the Sun.

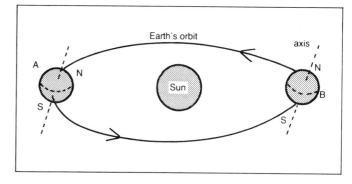

(a) When the Earth is at A which hemisphere will be having winter? Why?
(b) How long does the Earth take to revolve once round the Sun?
(c) Convert answer (b) into seconds.
(d) Assume that the orbit is circular. The radius of the orbit is 1.5×10^{11} m. Show that the speed of the Earth is about 30 km/s.

20 (a) How long does the Moon take to orbit the Earth?
  (b) Why does the Moon always show the same side towards the Earth?
  (c) Why does the Earth have day and night?
  (d) What are the positions of the Sun, Earth and Moon to give a spring tide?
  (e) What is the name often give to smaller tides?

21 The figure shows light from the Sun being blocked by the Moon.

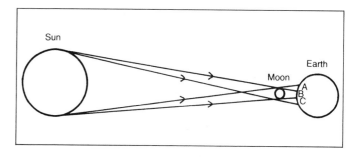

(a) What is the name given to this event?
(b) Use sketches to help you describe what will be seen at positions A, B and C.
(c) What is the name given to total darkness?
(d) Why does this event not happen every month?

Refraction and colour

How is light refracted?

- A little of the light is **reflected** from the surface of the glass. (You can often see reflections in a glass window.)
- Most of the light passes through the glass. This **transmitted** light lets you see through a glass window.
- Some of the light is **absorbed** as it passes through the glass. Coloured glass absorbs some colours more than others.

As light enters the glass it changes direction. This *bending* of light rays is called refraction.

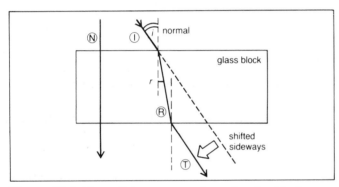

Figure 15.1 *How a light ray passes through a glass block:*

Ⓘ = incident ray; Ⓡ = refracted ray; Ⓣ = transmitted ray; *i* = angle of incidence; *r* = angle of refraction.

The normal is a line drawn at right angles to a surface where a light ray enters or leaves.

1. All angles are measured between a ray and the normal.
2. Refraction occurs both when the ray enters the block and when it leaves.
3. The light ray is *bent towards the normal* as it *enters* the glass block.
4. When it *leaves* the glass the ray is *bent away from the normal*.
5. When the ray enters the glass the angle of refraction *r* is smaller than the angle of incidence *i*. When it leaves *r* is greater than *i*.
6. The transmitted ray Ⓣ is parallel to the incident ray Ⓘ. But it is shifted sideways.
7. A ray may enter along a normal Ⓝ (i.e. at right angles to the surface). Then it is *not* bent. In this case angle *i* = angle *r* = 0.

How do real and apparent depths differ?

Measuring the real and apparent depth of a glass block

To find the apparent depth, the block is stood on end on a sheet of paper. A line drawn on the paper is seen through the top of the block. A search pin moved up and down the side of the block is used to point to the apparent position of the line. Both **real** and **apparent depths** of the block are measured down from the top.

- This effect is caused by light moving more slowly in glass than in air.

$$\frac{\text{speed of light in air}}{\text{speed of light in glass}} = \frac{\text{real depth}}{\text{apparent depth}} = \text{refractive index}$$

This ratio is called the **refractive index.** It has no units. It is about 3/2 for glass.

The speed of light in air is 3×10^8 m/s.
In glass it is 2×10^8 m/s.

Example: A swimming pool appears to be 1.5 m deep. The refractive index of water is 4/3. Find the real depth of water in the pool.

$$\text{refractive index} = \frac{\text{real depth}}{\text{apparent depth}}$$

$$\therefore \quad \frac{4}{3} = \frac{\text{real depth}}{1.5\,\text{m}}$$

So the real depth is: $\frac{4}{3} \times 1.5\,\text{m} = 2.0\,\text{m}$

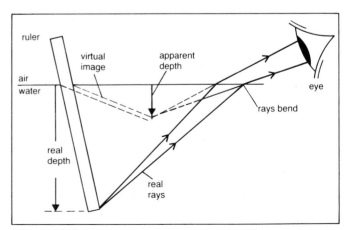

Figure 15.2 *How water appears to bend a ruler:*

The light rays from the bottom of the ruler are bent as they leave the water. They are bent away from the normal. This makes the ruler appear bent and shorter.

Prisms and colour

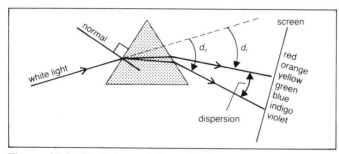

Figure 15.3 *Deviation and dispersion by a prism:*

Deviation is the change in direction of a light ray caused by the prism.

d_v = deviation of violet light. It is bent most.
d_r = deviation of red light. It is bent least.

Dispersion is the splitting of white light into the colours of the spectrum. The colours are not in stripes but change gradually through many different shades of colour.

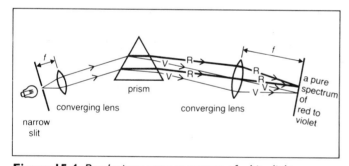

Figure 15.4 *Producing a pure spectrum of white light:*

To get a *pure* spectrum of light there must be no mixing or overlapping of the colours. To do this we need a parallel beam of white light. This is produced by a converging lens. The parallel beam then passes through a prism. This gives a parallel beam of each colour or part of the spectrum. Another converging lens then focuses each beam separately on to a screen.

The three **primary colours** of light are those which cannot be made by adding (or mixing) any other colours of light together. The primary colours of light are: red, green and blue.

The three **secondary colours** of light are those made by adding two primary colours together.

Primary colours of light added	Secondary colour made
Red + green	Yellow
Red + blue	Magenta
Blue + green	Cyan

	Primary + secondary = white
Complementary pairs add to make white light:	Blue + yellow = white Red + cyan = white Green + magenta = white
The three primary colours add to make white light:	Blue + red + green = white

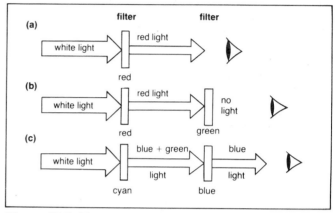

Figure 15.5 *Filters absorb certain colours of light:*

(a) The red filter absorbs all colours except red. So it transmits only red light.
(b) The eye sees black. This is because the green filter absorbs the red light.
(c) Blue is the only colour which both filters transmit. So the eye sees blue.

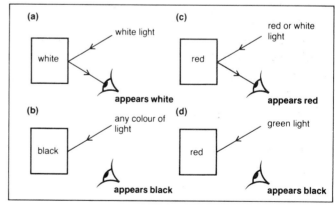

Figure 15.6 *Coloured surfaces reflect light:*

(a) White surfaces reflect all colours of light and appear white.
(b) Black surfaces reflect no colours. They absorb all colour and appear black.
(c) Red surfaces reflect red light. They absorb all colours except red and appear red.
(d) Red surfaces absorb the green light and so appear black.

16

Reflection

Mirrors

Plane mirrors and *polished surfaces* give **regular** reflections. All parallel rays are reflected in the *same* direction. Images or 'reflections' can be seen behind the surface. *Rough surfaces* like paper give **diffuse** reflections. All rays are reflected in *random* directions. No images are formed. Surfaces have a 'matt' look.

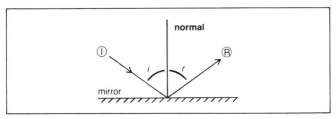

Figure 16.1 *The angles of incidence and reflection:*

Ⓘ = incident ray; Ⓡ = reflected ray; i = angle of incidence; r = angle of reflection.

The normal is a line drawn at right angles to the mirror where an incident ray strikes it.

The laws of reflection:

● The angles of incidence and reflection are equal.
● The incident ray, the reflected ray and the normal all lie in the same plane (flat surface).

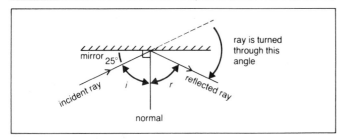

Figure 16.2:

Example: The angle between an incident ray and a plane mirror is 25°. Find (a) the angle of incidence, (b) the angle of reflection and (c) the angle through which the light ray is turned.
(a) The angle of incidence:
$$i = 90° - 25° = 65°$$
(b) The angle of reflection:
$$r = \text{angle } i = 65°$$
(c) The ray is turned through:
$$180° - (i + r)$$
$$= 180° - (65° + 65°) = 50°$$

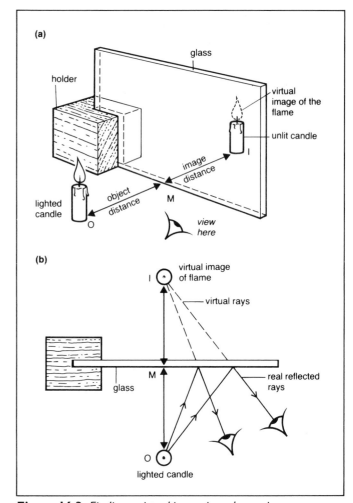

Figure 16.3 *Finding a virtual image in a plane mirror:*

(a) A sheet of glass is held upright. A lighted candle is then placed in front of it. Look through the sheet of glass. You can 'see' an image of a lit candle. Keep looking through the glass and place an unlit candle in the same position as the image of the lighted candle. The unlit candle will appear to burn! The object and image distances will be equal.

(b) The ray diagram shows how the virtual image is formed. The virtual rays show where the reflected rays appear to come from. The front of the glass acts like a plane mirror.

Properties of images formed in plane mirrors

● The image is the **same size** as the object.
● The image is **virtual** (meaning 'not really there', as in an illusion).
● The image is the same distance behind the mirror as the object is in front. Both lie on a line at right angles to the mirror. (In figure 16.3 IM = MO.)
● The image is *upright*. But the left and right sides are swapped round. (This is called **lateral inversion**.)
Try reading a clock reflected in a mirror.

When does total internal reflection occur?

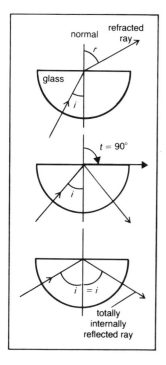

Figure 16.4 *Critical angle:*

i < c: when the angle of incidence *i* is smaller than the critical angle *c*, the light is refracted. The angle of refraction *r* is greater than angle *i*.

i = c: the **critical angle.** The angle *r* is 90°. The refracted ray grazes along the outside surface of the glass. This angle of incidence is called the critical angle *c*. From glass to air *c* = 42°.

i > c: **total internal reflection.** When the angle *i* is greater than the critical angle *c*, no light can get out of the glass by refraction. This is because the angle *r* would have to be greater than 90°. The inside surface of the glass acts as a perfect mirror.

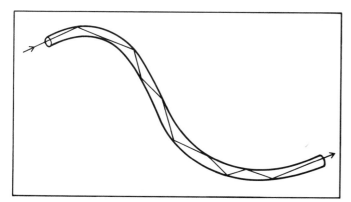

Figure 16.5 *An optical fibre or light pipe:*

An **optical fibre** is a long glass rod which is only a fraction of a millimetre thick. So it is quite flexible. This fibre of solid glass can be used as a **light pipe.** This is because rays of light travelling along inside it are totally internally reflected from the inside of the fibre's surface. So they cannot escape.
- Optical fibres can carry light round bends. This allows a doctor to see inside your body.
- Optical fibres can also carry information in the form of a digital code of light pulses. They carry telephone messages and computer data.

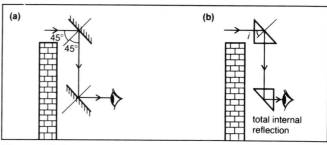

Figure 16.6 *Periscopes can use either plane mirrors or right-angled prisms:*

Both mirrors (a) and prisms (b) are mounted inside a tube.

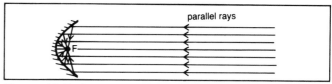

Figure 16.7 *Concave mirrors or reflectors:*

Parallel rays are converged (brought together) by a concave mirror to pass through a single point called the principal focus (F).

- A **parabolic** shape gives the best point focus.
- A radio telescope uses such a metal reflector. This focuses weak radio signals from space on to an aerial placed at the principal focus.
- Turn all the arrows round. You will see that the rays in the diagram can travel in the opposite direction. The rays, spreading out from the focus, are reflected into a parallel beam of light or microwave radiation.

Examples

Torch lamp, car headlamp, microwave communication dish, electric fire reflector.

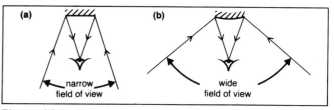

Figure 16.8 *Comparing a convex and plane mirror:*

(a) A plane mirror gives a narrow field or angle of view. But the images are their normal size. An example is a car rearview mirror. The driver gets a correct idea of how far the traffic is behind him.

(b) A convex mirror of the same size gives a *wider* field of view. But the images are smaller. An example is a car convex wing mirror. They are also used in shops to watch for shoplifters.

17 Lenses

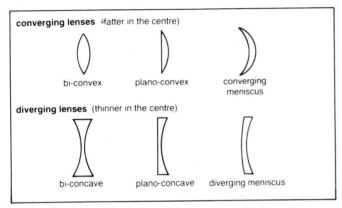

Figure 17.1 *Lens shapes:*

Converging lenses are fatter in the centre. They *converge* or bring together light rays.

Diverging lenses are thinner in the centre. They *diverge* or spread out light rays.

Lens definitions

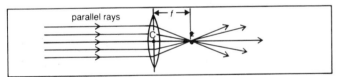

Figure 17.2 *The principal focus F and focal length f of a converging lens.*

The centre of a lens is known as its optical centre C.

The principal axis is the line through the optical centre drawn at 90° to the lens.

The principal focus F of a converging lens is the point to which all rays arriving parallel to the principal axis converge after refraction by the lens. Because the light actually passes through the focus it is called a *real* focus.

The focal length *f* of a lens is the distance between its optical centre and principal focus.

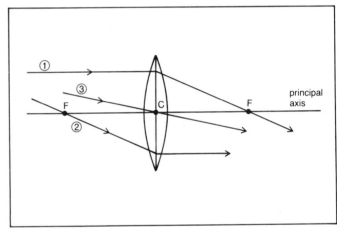

Figure 17.3 *Three special rays used in lens ray diagrams:*

① A ray parallel to the principal axis is refracted to pass through F.

② A ray arriving through F is refracted parallel to the principal axis (① reversed).

③ A ray through the centre C is undeviated.

(a) Converging lens (O between F and 2F)
e.g. (i) projector
     (ii) microscope objective lens

(b) Converging lens (O beyond 2F)
e.g. (i) camera
     (ii) the eye

(c) Diverging lens (O anywhere)
e.g. (i) eyepiece in some instruments
     (ii) in spectacles to correct short-sightedness

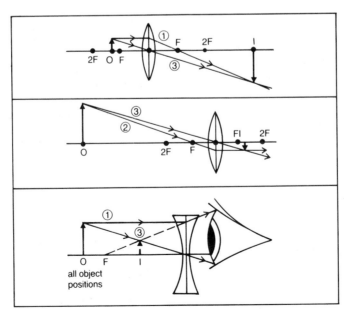

Image I is:
(i) real (can be formed on a screen)
(ii) inverted
(iii) magnified
(iv) on opposite side of lens to O, beyond 2F

(i) real
(ii) inverted
(iii) diminished
(iv) on opposite side of lens, between F and 2F
(This is diagram (a) reversed.)

(i) virtual
(ii) erect
(iii) diminished
(iv) on same side of lens as O, but nearer

Figure 17.4 *How images are formed by lenses:*

How can we measure the focal length of a lens?

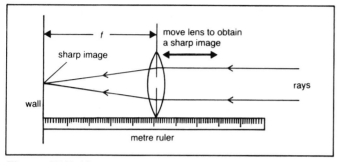

Figure 17.5 *Measuring the focal length of a converging lens:*

The rays of light from a distant object arrive at the lens nearly parallel. So the image will be formed very near the principal focus. Use a bright object such as a window at the far end of the room to form a *sharp* image on a white wall or screen. The distance from the sharp image to the lens is the focal length f.

What is the power of a lens?

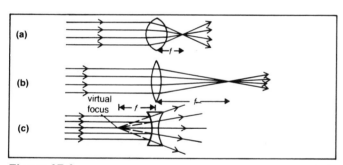

Figure 17.6:

(a) A powerful converging lens refracts (bends) the rays a lot. It is fat at the centre. The focal length f is short.

(b) A weak converging lens has a long focal length f and is much flatter.

(c) A powerful diverging lens is thick at the edges and is thin in the centre. The focal length f is short.

$$\text{power} = \frac{1}{\text{focal length in metres}} = \frac{1}{f}$$

- The power of a lens is measured in **dioptres (D)**.
- Converging lenses have *positive* powers.
- Diverging lenses have *negative* powers.

Example: A converging (convex) lens has a focal length of 20 cm. What is its power?
In metres: $f = 0.20$ m

$$\therefore \text{power} = \frac{1}{f} = \frac{1}{0.20} = +5 \text{ dioptres}$$

A prescription for spectacles gives the power of the lenses as -4.0 D. What kind of lens will be fitted? What is its focal length?

The $-$ power means that the lenses are diverging. They are thinner in the middle. Spectacles also use meniscus-shaped lenses.

$$\text{focal length} = \frac{1}{\text{power}} = \frac{1}{-4.0} = -0.25 \text{ metres}$$

The eye

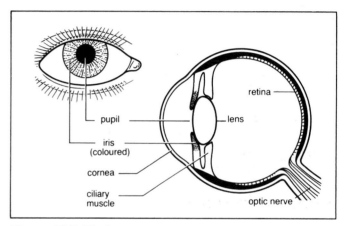

Figure 17.7 *The human eye:*

The **cornea** is a transparent 'window'.

The **retina** contains cells that convert light into electrical signals. These are sent down the optic nerve to the brain.

The **lens** changes shape to focus near and far objects.

The **iris** is the coloured circular ring of muscle in front of the lens. It controls the amount of light allowed in. This is done by varying the size of the pupil.

The **pupil** is the black hole in the centre of the iris. Light enters through it. This hole appears black except when lit by flash light.

The **ciliary muscle** is a ring of muscle. By contracting or relaxing, it changes the shape of the lens to allow focusing. When the ciliary muscle contracts, the lens becomes fatter and more powerful. Then it can focus near objects.

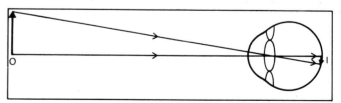

Figure 17.8 *How does the eye form an image?*

Rays from a large object (O) pass straight through the centre of the eye lens. They cross over to form a real image on the retina. This image (I) is *upside-down* and *diminished* (smaller).

Optical instruments

How does a magnifying glass work?

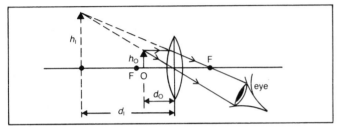

Figure 18.1 *The magnifying glass:*

This magnifies things close up. The magnifying glass is a convex lens with the object O *closer* to the lens than its principal focus F. The image I is virtual (not really there). It is *upright* and *magnified*. It can be seen only by looking through the lens.

● Use broken lines for virtual rays and images.

Magnification

● $\qquad$ magnification $= \dfrac{\text{height of image}}{\text{height of object}} = \dfrac{h_I}{h_O}$

By similar triangles:

$\qquad$ magnification $= \dfrac{h_I}{h_O} = \dfrac{d_I}{d_O} = \dfrac{\text{image distance}}{\text{object distance}}$

● Bringing the object closer to the lens increases the magnification.

The projector or enlarger

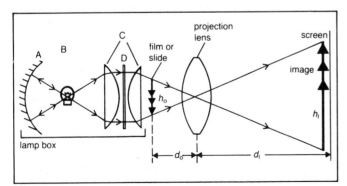

Figure 18.2 *The projector or enlarger magnifies small pictures on film:*

● The **lamp box** gives even, bright illumination of the object. It has four parts:
A is the concave reflector. It reflects light forward.
B is the lamp.
C is the condenser lens. It concentrates the light evenly on to the film or slide.
D is the filter. It absorbs heat.
● The **film** or **slide** forms the bright object for the projection lens. It should be upside-down for an upright image.
● The **projection lens** is moved in or out of the projector. This focuses the image on the screen.
● The **image** is real, upright and magnified.
By similar triangles:

$$\text{magnification} = \frac{h_I}{h_O} = \frac{d_I}{d_O}$$

$$= \frac{\text{image distance}}{\text{object distance}}$$

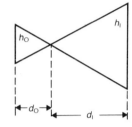

Moving the lens closer to the object increases the magnification. Moving the screen further away from the lens does the same.

What is a pinhole camera?

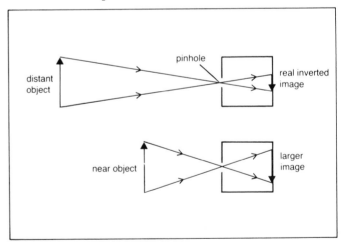

Figure 18.3 *The pinhole camera:*

This is a light-tight box with a small pin-sized hole in the front. It can form sharp images of both near and far objects. The size of the image gets larger as the object gets nearer the pinhole.

● If there are several pinholes in the camera, each pinhole produces an image in a slightly different position. The image becomes more blurred as a pinhole is enlarged.
● A lens is placed in front of the multiple pinholes or the enlarged hole will collect all the light rays. It will then focus them into a single sharp image as in a lens camera.

The lens camera

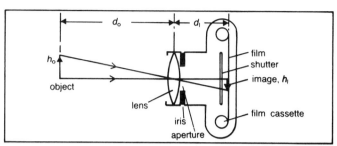

Figure 18.4 *The lens camera:*

- The **image** is real, upside-down and diminished.
- The **shutter** is a spring-loaded metal or fabric blind. It covers the film except for the moment when the photograph is taken.
- The **lens** is moved in or out to focus the image.
- The **iris** is a multi-leaved metal structure. It forms the hole or aperture through which light enters.
- The **aperture** is the size of the hole in the centre of the iris. A small aperture is used when the light is bright.

Magnification

$$\text{magnification} = \frac{h_I}{h_O} = \frac{d_I}{d_O} = \frac{\text{image distance}}{\text{object distance}}$$

The magnification of the image increases if:
- the object is nearer to the camera;
- the lens has a longer focal length. This increases d_I. (An example is a telephoto lens.)

Controlling the light

The aperture controls how much light reaches the film. The size of the aperture is given in steps called **stops**. For each stop the area and the light entering the camera halves. The stops are numbered with an **f-number**.

$$\text{f-number} = \frac{\text{focal length of lens}}{\text{diameter of aperture}}$$

A/4	A/8	A/16	A/32	A/64	area
◯	◯	◯	○	○	
12.5	9	6.25	4.5	3.1	diameter in mm
4	5.6	8	11	16	f number

Figure 18.5 *f-numbers or stops for a lens of focal length 50 mm.*

Shutter speed

When the photograph is taken the **shutter** opens. This exposes the film for a short time. The exposure time or shutter speed determines how much light reaches the film.

Short exposures are needed to 'freeze' moving objects.
- 1/(speed number) gives the exposure time in seconds. For example number 500 gives an exposure of 1/500 or 0.002 seconds.

How do glasses correct faulty eyes?

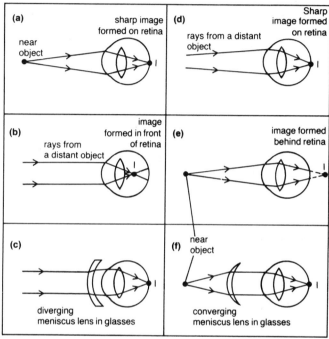

Figure 18.6 *Correcting short sight (myopia):*

Figure 18.7 *Correcting long sight (hypermetropia):*

(a) A **short-sighted** or **myopic eye** can see near objects clearly. This is because the eye focuses the light to form a sharp image, I, on the retina.

(b) The same eye cannot see distant objects clearly. This is because it brings the rays together in front of the retina at I. The image on the retina is blurred.

- This defect may be caused by too great a curvature of the cornea. Or the eyeball may be too long. Many young people are short-sighted.
- Short sight is corrected (c) by using glasses with a *diverging* lens. This opens out the rays so that they enter the eye as if they came from a near object. They are then focused on the retina.

(d) A **long-sighted eye** can see clearly only objects which are quite far away. When this is greater than arm's length away the person has difficulty in reading.

(e) Rays from a near object are bent too little and so meet beyond the retina at I. The image on the retina is blurred. *Think of a possible cause for this defect.*

- A *converging* lens (f) corrects the defect by bending the rays inwards. They then appear to come from a more distant object and meet on the retina. This forms a sharp image.

Review questions: Chapters 15, 16, 17 and 18

C15 1 The figure shows a light ray passing through a glass block. Which ray is the:
(a) incident;
(b) transmitted;
(c) reflected;
(d) refracted?

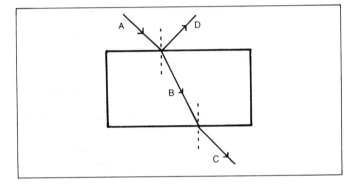

2 Copy and complete the following:
(a) Refraction occurs when a ray enters a glass block and when it _____.
(b) A line drawn at _____ _____ to a surface where a ray enters or leaves is called a normal.
(c) The light ray bends _____ the normal as it enters the block and _____ _____ the normal as it leaves.
(d) All angles are measured between a _____ and a normal.

3 A boy watches a fish swim. He thinks it seems to be at a depth of 60 cm.
(a) The refractive index of water is 4/3. What is the real depth of the fish?
(b) What is this effect caused by?

4 The figure shows white light passing through a glass prism.
(a) Which colour is bent most?
(b) What are the angles A and B called?
(c) What is angle C called?

5 (a) Name the three secondary colours.
(b) Red light + blue light = _____.
(c) Red light + cyan light = _____.
(d) Alison wears a red dress to a disco. What colour will it look in green light?
(e) Describe how a white flower with green leaves in a blue pot looks in: (i) red light; (ii) yellow light.
(f) What colour will be seen after white light has passed through a yellow filter and then a green filter?

C16 6 Copy and complete the following:
(a) For a *polished* surface all parallel light rays are reflected in the _____ direction.
(b) A *rough* surface has a matt look. This is because light rays are reflected in _____ directions.
(c) An incident ray, reflected ray and the normal all lie in the _____ plane.

7 If the angle between an incident ray and a plane mirror is 40°, calculate:
(a) the angle of incidence;
(b) the angle of relection;
(c) the angle through which the light ray is turned.

8 Which of the following are properties of an image in a plane mirror?
(a) virtual: (b) real; (c) inverted;
(d) enlarged; (e) laterally inverted.

9 Suppose you held this page up to a plane mirror. Sketch what the following symbols would look like:
(a) \ (b) <<<<—— (c) [×<v}

10 An optical fibre is a thin glass fibre which carries light rays.
(a) Are the angles of incidence of the light rays greater than, the same as, or smaller than the critical angle?
(b) What is this kind of reflection called?

11 (a) In a periscope using two parallel plane mirrors, what angle does each mirror make with the incident light ray?
(b) What angle does each mirror turn the light ray through?
(c) What is often used to replace the mirrors in the periscope?

12 The figure shows a selection of mirrors.

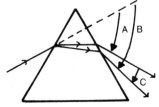

(a) Name the four mirrors shown in the figure.
(b) Which shape of mirror could be used as:
(i) a car wing mirror;
(ii) a radio telescope dish?

C17 13 Copy and complete the following:
  (a) A *diverging* lens is _____ in the centre.
  (b) A *convex* lens _____ the light rays.
  (c) The middle of a lens is called the _____.
  (d) The focal length of a lens is the _____ from
      its _____ to the principal focus.

14 The figure shows a ray diagram for a convex lens.

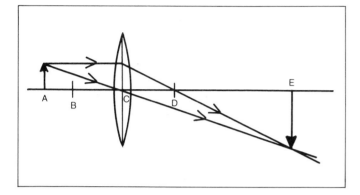

Use the letters to describe the:
(a) object;   (b) image;   (c) principal focus;
(d) focal length;   (e) optical centre.

15 (a) Which will be *more* powerful, a lens with long
       or short focal length?
    (b) What are the units of power for lenses?
    (c) What type of lenses have *negative* powers?
    (d) What units must the focal length be in when
        calculating the power of a lens?

16 (a) What is the power of a lens of focal length: (i)
       25 cm; (ii) 50 cm; (iii) − 10 cm?
    (b) A prescription for spectacles gives the power
        as + 2.0 D. What kind of lens will be fitted?
        What is its focal length?

17 The figure shows an eye. Which letter indicates the
    following:
    (a) lens;
    (b) ciliary muscle;
    (c) cornea;
    (d) retina;
    (e) optic nerve;
    (f) pupil?

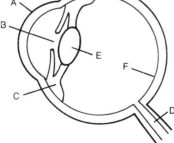

18 (a) Give three properties of an image formed in
       the eye.
    (b) Where is the image formed in the eye?
    (c) What happens to the shape of the lens when
        the ciliary muscles relax?
    (d) What happens to the power of the lens when
        the eye focuses on a near object?

(e) Name the coloured part of the eye.
(f) What is the pupil for?

C18 19 (a) What kind of lens is used for a magnifying glass?
    (b) Give three properties of an image formed in a
        magnifying glass.
    (c) What happens to the magnification when you
        bring the object closer to the lens?
    (d) An image is four times bigger than the object.
        What is the magnification?

20 (a) Name three properties of an image formed in a
       projector.
    (b) What is the purpose of the condenser lens?
    (c) Which way does the screen need to be moved
        to make the image bigger?

21 Copy and complete the following:
    (a) The image size increases when a pinhole
        camera is moved _____ the object.
    (b) Four small pinholes would give _____ sharp
        images which may overlap.
    (c) A large pinhole gives a _____ image.

22 Julia wishes to take a photograph of a motor bike
    race.
    (a) What sort of exposure time should she use to
        'freeze' the motion of the bikes?
    (b) When should she use a larger aperture?
    (c) What are the aperture settings called?
    (d) What are the aperture step sizes called?

23 A lens camera with a focal length of 50 mm is set to
    an f-number of 8. The diameter of the aperture is
    6.25 mm.
    (a) What will be the size of the aperture if the
        f-number is changed to 4?
    (b) What will be the f-number if the aperture
        changes to 3.1 mm?

24 The figure shows eyes with defects.

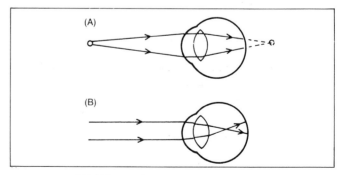

(a) What is the defect in the eye in (A)?
(b) Which lens would correct this defect?
(c) Where is the image formed in (B)?
(d) Can this eye see a ship at sea clearly without
    spectacles?

Waves

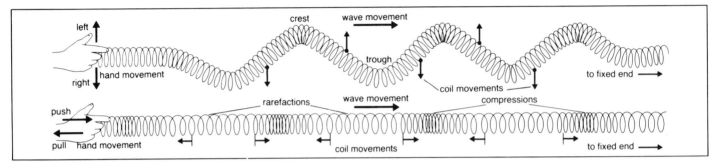

Figure 19.1 *Waves on a slinky spring:*

The wave motion moves along the spring. As it travels the coils are displaced (shifted) from their usual or resting positions (shown by the small arrows). (a) A transverse wavetrain. (b) A longitudinal wavetrain.

In transverse waves the displacement of the particles (coils on the spring) is at *right angles* to the direction of travel of the wave motion.

In longitudinal waves the displacement of the particles is *forwards and backwards* along the direction of travel of the wave motion.

Transverse waves	Longitudinal waves
Have **crests** and **troughs**	Have **compressions** and **rarefactions**
Examples include: most water waves, waves on a string, light waves and all other electro-magnetic waves	Examples include: dominoes toppling, wagons shunting along a train, sound waves in air, pipes and wind instruments

Waves can also be classified another way:

Mechanical waves	Electromagnetic waves
Need a medium or material to travel through Cause individual particles in the medium to vibrate	Do not need a medium to travel through Can travel through a vacuum and through space
Examples include: sound waves, water waves, shock waves including earthquakes	Examples include: γ-rays and X-rays, UV light, IR radiation, microwaves and radio waves

What are travelling or progressive waves?

A travelling wave carries energy with it.

- The medium or material through which a wave travels does not travel with the wave.
- The particles of the medium are displaced by the wave. They vibrate about rest positions.
- The shape of a wave stays the same as it travels. But its amplitude gets smaller as energy is lost or the waves spread out.

Examples

- Water waves carry energy to toss ships, wash away cliffs or drive wave-energy generators.
- Light waves carry energy. This can be converted into electrical energy by solar cells.

What are stationary or standing waves?

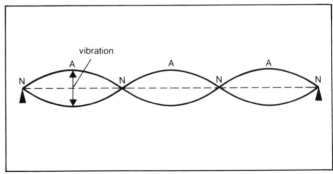

Figure 19.2 *Stationary or standing waves on a string:*

Standing waves are formed when a travelling wave is reflected back along the same path and overlaps itself.

- The wave motion appears to stand still and neither moves itself nor carries energy along.
- There are fixed positions on standing waves, called **nodes** (N). Here the amplitude is zero.
- The amplitude is maximum at the **antinodes** (A).

Examples

- Waves on stringed instruments.
- Waves inside pipes or wind instruments.
- Light waves inside a laser. They are reflected back between parallel mirrors.

Graphs of waves

Suppose we were to take a photograph of a transverse wavetrain (e.g. waves on water) from a side view (profile). We would have a **displacement–position** graph like figure 19.3 (a).

Suppose we watch a cork floating on the water at points A or C along the wave. As the wave moves to the right we see the cork bob up and down, as shown in figure 19.3 (b). This is a **displacement–time** graph. It shows the movement of the cork (or any single particle) in the wave at points A and C. The cork (or particle) is **oscillating** up and down.

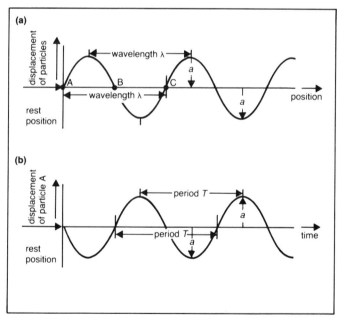

Figure 19.3 Graphs of wavetrains: (a) displacement–position graph; (b) displacement–time graph.

> **The wavelength λ of a transverse wave is the distance between two successive crests or two successive troughs.**

λ is measured in metres. (Successive = next to each other.)

Particles at points A and C are one wavelength apart. They have the same displacement and are moving in the same direction at the same time.

- The particle at B has the same displacement as those at A and C. But it is moving in the opposite direction.
- From A to B is half a wavelength.

> **The amplitude *a* of a wave is the *maximum displacement* of the particles in the wavetrain from their rest positions.**

The full height of a wave from the bottom of a trough to the top of a crest is *twice* the amplitude.

> **The period *T* is the time a particle in the wavetrain takes to make one complete oscillation.**
>
> **The frequency *f* is the number of complete oscillations made in 1 second by a particle in the wavetrain.**

T is measured in seconds. *f* is measured in hertz (Hz).

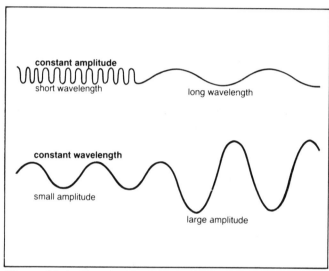

Figure 19.4 *Changing the wavelength or amplitude.*

The wave equation

The **speed** of a wave *c*:
- does *not* depend on its shape, amplitude, frequency or wavelength;
- *does* depend on the nature of the material it travels through;
- can be calculated using the wave equation:

$$\text{wave speed} = \text{frequency} \times \text{wavelength}$$
$$c = f\lambda$$

Example: A cork bobs up and down on water as a wavetrain passes by. The crests of the waves are 2.0 metres apart. The cork completes four oscillations every 10 seconds. Calculate the wave speed.

$$\text{wavelength } \lambda = 2.0\,\text{m}$$
$$\text{frequency } f = 0.4\,\text{Hz (oscillations in 1 second)}$$
$$\therefore \text{ wave speed } c = f\lambda = 0.4 \times 2.0 = 0.8\,\text{m/s}$$

20

Watching waves

Waves can be generated on a ripple tank by the shaky movement of an electric motor. This has a weight attached off-centre to its axle.

● *Straight* or *plane waves* are produced when the beam holding the motor touches the water.
● *Circular* waves are produced when a small pea-sized dipper touches the water.

How are waves reflected?

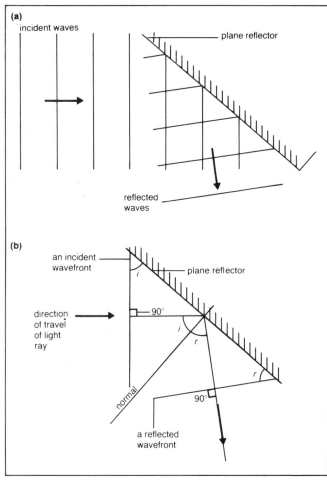

Figure 20.1 *Reflection of straight waves by a plane reflector: The reflection of waves (a) and of a light ray (b) obey the same laws:*

(a) Both incident and reflected wavefronts are straight and have equal spacings. The incident and reflected waves have the same speed and wavelength.

(b) The angle of incidence i is equal to the angle of reflection r at all angles. The incident and reflected wavefronts are at right angles to their direction of travel.

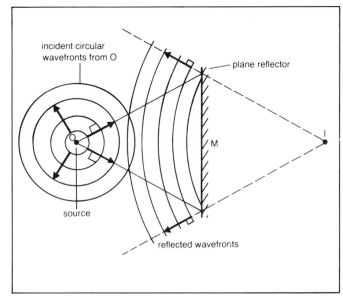

Figure 20.2 *Circular waves reflected by a plane reflector:*

When circular waves are reflected by a plane reflector the reflected waves are also circular. They appear to come from a point I behind the reflector. I is also where the *virtual* image formed by a plane reflector would be found. OM = MI. OI is at 90° to the reflector.

How are waves refracted?

Refraction is caused by a change of wave speed when a wave passes from one medium to another.

Waves travel more slowly in shallow water. This allows us to study the refraction of waves as they pass from deep to shallow water.

$$\text{refractive index} = \frac{\text{speed of waves in first medium}}{\text{speed of waves in second medium}}$$

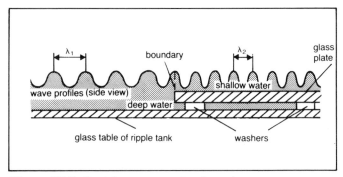

Figure 20.3 *Straight waves passing from deep water to shallow water:*

The frequency of the waves does not change. Their wavelength, however, shortens in proportion with their speed.

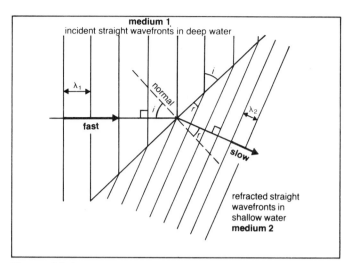

Figure 20.4 *Refraction of straight wavefronts at a plane boundary:*

The boundary is where waves pass from deep water to shallow water. When the waves meet a boundary at an angle their change of *speed* causes a change of *direction*. The waves are refracted.

How are waves diffracted?

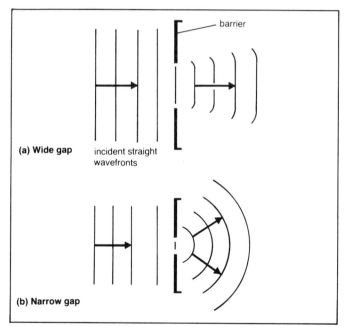

Figure 20.5 *Diffraction of straight wavefronts:*

When a wave meets a gap in a barrier, as it passes through, the shape of its wavefront changes.

(a) After a wide gap the wavefronts are still straight except for a slight edge curvature. There is almost straight-line travel.

(b) After a narrow gap circular wavefronts are produced. The waves spread out in all directions.

The spreading of waves round corners and edges of barriers is called diffraction.

● The spreading or diffraction of the waves is greatest when the gap is similar in size to the wavelength of the waves.

How do waves interfere?

Interference is the name given to the effects which occur when two separate wave motions overlap.

Waves do not bump into one another. Rather they pass through each other and merge or combine their effects.

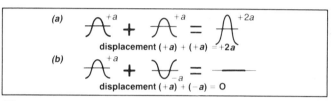

Figure 20.6 *Superposition of waves:*

The displacement of any particle caused by overlapping waves is the sum of the separate displacements caused by each wave at a particular moment.

(a) **constructive interference** occurs when the displacements due to two waves combine to produce a larger displacement. This happens when the wave crests overlap. The waves are said to be **in phase.**

(b) **Destructive interference** occurs when, in effect, the crest of one wave fills in the trough of another. This produces an effect of no displacement. Such waves are said to be **in antiphase.**

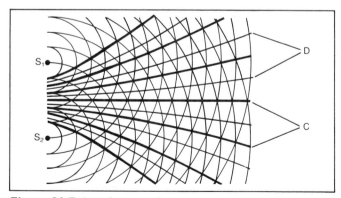

Figure 20.7 *Interference of waves:*

By attaching two dippers to the vibrating beam of a ripple tank, or by sending straight wavefronts through two gaps in a barrier, we can then see the effects of interference.

● Along the thick lines, marked C, the waves are in phase. They interfere constructively.

● Along the thin lines, marked D, the waves are in antiphase. They interfere destructively.

21 Sound

Sounds are made when things vibrate. Examples are the vocal chords in your throat, the taut strings on a guitar, the column of air in a pipe, the cone of a loudspeaker and the skin on a drum. Sound waves reach our ears by *longitudinal vibrations* of air molecules between the source and our ear drums. Sound waves are *mechanical waves*. They require a medium to travel through.

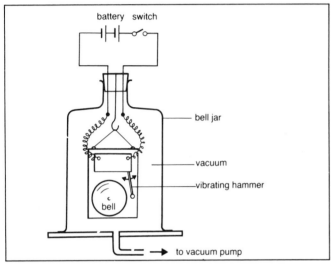

Figure 21.1 *Vibrations in a vacuum can be seen – but not heard!*

Air is pumped out of the bell jar. The hammer still strikes the bell. However, no sound can reach us. This is because sound cannot travel through a vacuum.

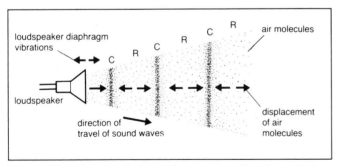

Figure 21.2 *Longitudinal sound waves travelling through air:*

The (invisible) air molecules are displaced backwards and forwards (**longitudinally**) along the direction of travel of the wave.

C = compression. Here the molecules are pushed closer together. The pressure increases.

R = rarefaction. Here the molecules are spaced out more thinly. The pressure decreases.

How can we describe sounds?

We can use an oscilloscope (CRO) to display the electrical oscillations produced by a signal generator or from a microphone. Then we are able to compare various sounds with their waveforms.

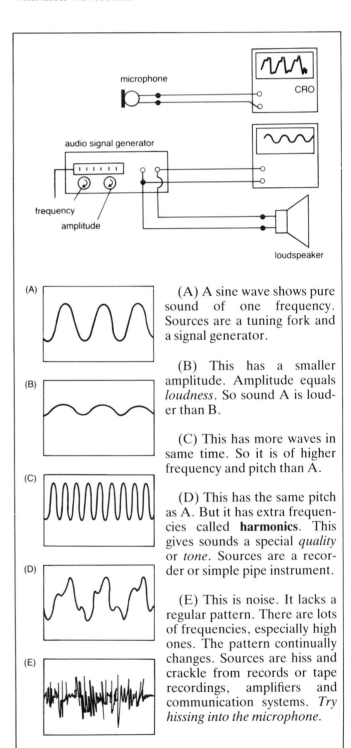

(A) A sine wave shows pure sound of one frequency. Sources are a tuning fork and a signal generator.

(B) This has a smaller amplitude. Amplitude equals *loudness*. So sound A is louder than B.

(C) This has more waves in same time. So it is of higher frequency and pitch than A.

(D) This has the same pitch as A. But it has extra frequencies called **harmonics**. This gives sounds a special *quality* or *tone*. Sources are a recorder or simple pipe instrument.

(E) This is noise. It lacks a regular pattern. There are lots of frequencies, especially high ones. The pattern continually changes. Sources are hiss and crackle from records or tape recordings, amplifiers and communication systems. *Try hissing into the microphone.*

Figure 21.3 *Hearing and 'seeing' sound waves:*

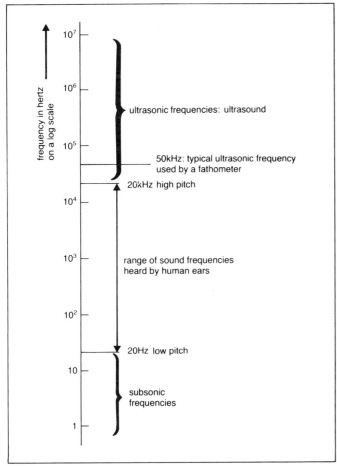

Figure 21.4 *The frequency spectrum of sound waves:*

Ultrasonic frequencies (1 MHz to 10 MHz) are used in medical diagnosis. Subsonic frequencies may be felt as vibrations or shock waves (e.g. earthquakes).

How can we measure the speed of sound?

To get an accurate value for the speed of sound we need to time a sound over a long distance.

Using an **echo method** doubles the distance the sound travels. Stand a measured distanced (say 100 m) from a large wall or building. Make sharp clapping sounds by banging together two blocks of wood. Repeat the claps at regular time intervals which coincide exactly with the echoes. Time 50 claps with a stop watch. Typical results for this experiment are:

$$\text{distance from wall} = 100 \text{ m}$$
$$\therefore \text{distance sound travels} = 200 \text{ m}$$
$$\text{time taken for 50 claps} = 30.3 \text{ s}$$
$$\therefore \text{time interval between claps} = 30.3/50 = 0.606 \text{ s}$$
$$\text{speed} = \frac{\text{distance travelled}}{\text{time taken}} = \frac{200 \text{ m}}{0.606 \text{ s}} = 330 \text{ m/s}$$

Example: *Echoes.* A ship, surrounded by fog, sounds its horn. If an echo is heard from a cliff 5.0 seconds later, how far away is it? The speed of sound in air is 330 m/s.

$$\text{distance} = \text{speed} \times \text{time} = 330 \times 5.0 = 1650 \text{ m}$$

This is the distance travelled by the sound to the cliff and back again.

$$\therefore \text{distance to the cliff} = \tfrac{1}{2} \times 1650 \text{ m} = 825 \text{ m}$$

How is sound reflected?

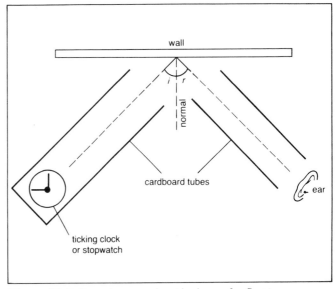

Figure 21.5 *Sound waves obey the laws of reflection:*

The reflected sound is the loudest when the angle of reflection r is equal to the angle of incidence i.

Reverberation

In a cathedral or large hall a sound is reflected around from many walls and lasts for a long time. At each reflection a little of the sound is absorbed. The reflected sound then becomes a little quieter. Many echoes may merge into one prolonged sound. This effect is known as **reverberation**. Sound-absorbing materials are used on the walls, floor and ceiling of a concert hall to reduce the reverberation. However, if sound is absorbed too much the hall will sound 'dead'. Voices and music will then appear weak or muffled.

Sound pipes

Sound waves can be totally internally reflected. This is like light inside an optical fibre. Speaking tubes are used on ships. A flexible hollow tube carries sounds in a doctor's stethoscope. The head sets used by passengers to listen to music on many aircraft are connected by sound tubes (rather than wires carrying electrical signals).

22

The electromagnetic spectrum

Figure 22.1 shows all the members of the **electromagnetic radiation** family and how they fit together in a *continuous* spectrum. The frequency increases upwards by ten times for each rung of the 'ladder'. This is a **log scale.**

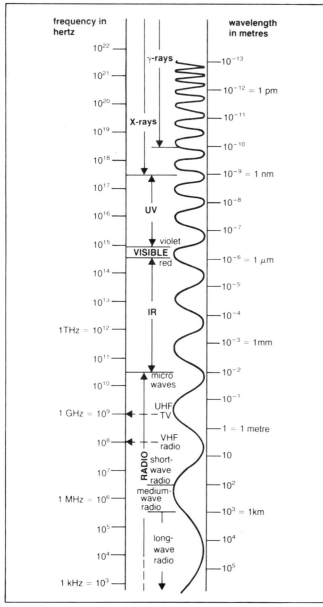

Figure 22.1 *The electromagnetic spectrum.*

Which properties are common to all parts of the spectrum?

All electromagnetic waves –
① Transfer energy from one place to another.
② Can be emitted and absorbed by matter.
③ Do *not* need a medium to travel through.
④ Travel at 3.0×10^8 m/s in a vacuum or through space.
⑤ Are transverse waves.
⑥ Obey the laws of reflection and refraction.
⑦ Are diffracted by openings or obstacles in their path.
⑧ Produce interference effects when waves overlap.
⑨ Carry no electric charge.

What are typical wavelengths and frequencies?

The range of wavelengths in the spectrum is wide. It goes from 1 picometre (10^{-12} m) for gamma rays (γ) up to kilometres for radio waves. You can think of things of similar dimensions to each wavelength in the spectrum. This will help you to recognise where each wavelength belongs. For example:

The wavelengths of γ-rays and X-rays are similar to the dimensions of molecules (10^{-9} m), atoms (10^{-10} m) and atomic nuclei (10^{-14} m). It is not surprising to find that γ-rays are emitted by decaying nuclei. X-rays can be used to study the structure of crystals.

Light waves have wavelengths of a few tenths of a micrometre (0.4 to 0.7 μm). Light can be diffracted by narrow slits or scratches which are only a micrometre wide.

The lengths of the aerials fitted to VHF radios are a quarter of the wavelength of the radio waves they are designed to detect. They are typically 75 cm.

Examples: Calculate the frequency of the radio waves detected by an aerial 75 cm long.

The aerial detects radio waves of wavelength 4×0.75 m $= 3.0$ m. The frequency of these radio waves is given by:

$$f = \frac{c}{\lambda} = \frac{3.0 \times 10^8 \,\text{m/s}}{3.0 \,\text{m}} = 1.0 \times 10^8 \,\text{Hz or } 100 \,\text{MHz}$$

Calculate the wavelength of the microwaves used in a microwave oven. Their frequency is 2450 MHz. The speed of electromagnetic radiation is 3.0×10^8 m/s.

$$\lambda = \frac{c}{f} = \frac{3.0 \times 10^8 \,\text{m/s}}{2450 \,\text{MHz}} = 0.12 \,\text{m or } 12 \,\text{cm}$$

Name, typical wavelength and source.	Detectors, properties and uses
γ-rays $\lambda = 1\,pm = 10^{-12}\,m$ Source: nuclei of radioactive atoms and c **Cobalt-60 source** Cobalt – 60 source	**Detected by:** photographic film and G-M tube **Properties:** very penetrating; very dangerous **Uses:** treatment of cancerous growths; γ-camera forms images of inside body and finds flaws in metals; sterilises equipment in hospitals and food industry
X-rays $\lambda = 100\,pm = 10^{-10}\,m$ Source: X-ray tubes **X-ray tube** X-ray tube	**Detected by:** photo film and fluorescent screen **Properties:** same as γ-rays **Uses:** treatment of skin disorders; X-ray radiography; study of crystal structures; inspection of welds in steel pipes
Ultraviolet (UV) light, $\lambda = 10\,nm = 10^{-8}\,m.$ Source: the Sun, very hot objects, arcs and sparks and mercury lamps **UV lamp** UV lamp	**Detected by:** photo film, photo cells and fluorescent chemicals **Properties:** absorbed by glass; causes sunburn; damages and kills living cells **Uses:** skin treatment; gives a suntan; makes clothes washed with powders look whiter; with certain inks helps detect forgeries.
Visible light green $\lambda = 0.6\,\mu m$ $= 0.6 \times 10^{-6}\,m$ Source: the Sun, hot objects, lamps and lasers **Sun** Sun	**Detected by:** eye, photo film and photo cells **Properties:** refracted by glass lenses and prisms; focused by the eye **Uses:** essential for photosynthesis and plant growth; in communication systems with lasers and optical fibres
Infrared (IR) light, $\lambda = 100\,\mu m = 10^{-4}\,m.$ Source: the Sun, warm and hot objects, e.g. fires and people **Person** Person	**Detected by:** special photo film, skin, semiconductor devices: LDR and photodiode **Properties:** causes heating when absorbed; makes skin feel warm **Uses:** radiators and fires emit IR radiation; finding buried warm bodies; IR satellite photos reveal diseased crops
Microwaves, $\lambda = 12\,cm.$ Source: microwave ovens and communication links **Microwave communication dish** microwave communications dish	**Detected by:** microwave-receiving aerials **Properties:** absorbed by water and fats in food and people; is therefore dangerous **Uses:** microwave communication links; microwave cooking
Radio $\lambda = 3\,m$ (VHF radio) $\lambda = 1.5\,km$ (long waves) Source: TV and radio aerials and transmitters **Radio** Radio	**Detected by:** metal aerials and tuned circuits **Properties:** induces alternating electric currents in metals, e.g. aerials **Uses:** radio, TV and satellite communications; radar detection of ships and aircraft; radio astronomy

Review questions: Chapters 19, 20, 21 and 22

C19 1   (a) What are the two types of wave motion called?
  (b) Name the waves in which the particles move at 90° to the wave direction.
  (c) Which type of waves have crests?
  (d) Which type of waves have compressions?

2   Which of the following are examples of transverse waves:
  (a) radio waves;   (b) sound waves in air;
  (c) waves on a string;   (d) light waves?

3   (a) Give another name for a progressive wave.
  (b) Why does the amplitude get smaller as the wave travels?
  (c) Do the particles of the medium travel with the wave?
  (d) Is the wave speed affected by the medium?

4   For a stationary wave:
  (a) What is another name for this wave?
  (b) What are the fixed positions called?
  (c) Where is the wave amplitude greatest?

5   The figure shows a displacement–position graph of a transverse wave. Use the letters:

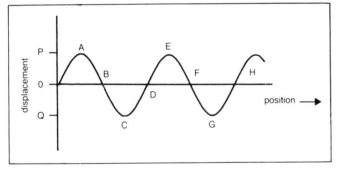

  (a) to indicate the wavelength of the wave with different pairs of letters;
  (b) to indicate the amplitude of the wave.

6   (a) What is the time to complete one oscillation of a wave called?
  (b) What is the number of oscillations per second called?
  (c) If a wave has ten oscillations per second what is the time for one oscillation?

7   A buoy bobs up and down on water. The crests of the waves are 2.0 m apart. The buoy completes 6 oscillations every 10 s. Find the speed of the wave.

C20 8   What attachment is used with a ripple tank to produce:
  (a) straight waves; (b) circular waves?
  Straight waves meet a straight barrier at an angle of incidence of 30°.
  (c) What is the angle of reflection?
  (d) Will the wavelength of the wave change?
  Circular waves meet a straight barrier 10 cm from their source.
  (e) Describe the reflected waves.

9   Copy and complete the following:
  (a) Refraction is caused by a change of wave _____.
  (b) When water waves move into a shallow bay their _____ and _____ decrease. But the _____ does not change.

10 (a) What is the spreading of waves round corners called?
  (b) What size gap gives most spreading of waves?
  (c) What happens when waves overlap?

11 The figure shows the superposition of waves.

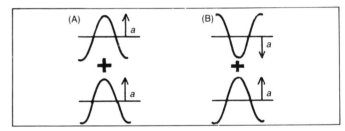

  (a) Describe, using diagrams, the result of the overlapping of these waves.
  (b) What are the new wave amplitudes?
  (c) Which diagram above shows constructive interference?
  (d) What is the phase difference in (B)?

12 The figure shows waves interfering. Assume the lines represent a wave crest of amplitude a. Copy and complete the table:

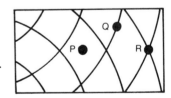

Point	Letter	Wave motion	Interference	Amplitude
Space + space	P	Trough + trough	Constructive	2a
Line + space				
Line + line				

58

C21 13 Copy and complete the following:
  (a) Sounds are made when things _____.
  (b) Sound waves are mechanical waves which need a _____ to travel through.
  (c) Sound waves do not travel in a _____.

14 The figure shows invisible air molecules in front of a vibrating loudspeaker cone.

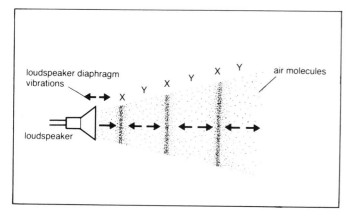

  (a) What is the name of this kind of wave?
  (b) What are the sections X and Y called?
  (c) How does the spacing of air molecules at X and Y differ?
  (d) Where is the pressure smallest?
  (e) What is the angle between the direction of travel of the sound waves and the displacement of the air molecules?

15 Anthony sings into a microphone connected to an oscilloscope. What happens to the waveform on the oscilloscope if:
  (a) he sings quietly;
  (b) he sings at a lower pitch;
  (c) he hisses?

16 Saeed stands 660 m from a cliff. She then shouts 'hello'. The speed of sound in air is 330 m/s. How long must she wait for the echo?

17 A ship measures the depth of the sea using ultrasound. The speed of ultrasound waves in sea water is 1500 m/s.
  (a) If the time between sending out a wave and receiving the echo is 2 s, calculate the distance the wave travelled.
  (b) What is the depth of the sea-bed?

18 What is the effect called when many echoes merge into one prolonged sound?

C22 19 The strip shows the electromagnetic spectrum.

A	X-rays	UV	B	IR	C	TV	D

  (a) What are regions A, B, C and D called?
  (b) Which is the high-frequency end of the spectrum?
  (c) What do the letters UV stand for?
  (d) Why is this spectrum called continuous?

20 Copy and complete the following about electromagnetic waves:
  (a) They transfer _____ from one place to another.
  (b) They show _____ effects when they overlap.
  (c) They are _____ longitudinal waves.
  (d) They all travel at a speed of _____ in a vacuum.
  (e) They do not need a _____ to travel through.

21 What type of electromagnetic radiation has a wavelength similar in size to:
  (a) the diameter of molecules;
  (b) a narrow slit (cut) made by a razor blade;
  (c) VHF radio aerials;
  (d) a ruler?

22 (a) What is the wavelength of:
    (i) orange light, frequency 5×10^{14} Hz;
    (ii) radio waves, frequency 3×10^6 Hz?
  (b) What is the frequency of:
    (i) gamma rays, wavelength 3 pm;
    (ii) IR, wavelength 1 mm?
    (Speed of waves, $c = 3 \times 10^8$ m/s.)

23 Calculate the frequency of radio waves detected by an aerial 1.5 m long.
  (Speed of waves, $c = 3 \times 10^8$ m/s.)

24 Which type(s) of electromagnetic radiation is:
  (a) emitted by humans;
  (b) absorbed by glass;
  (c) detected by Geiger–Müller tubes;
  (d) able to induce currents in metals;
  (e) absorbed by water and fats in food?

25 Which kind of radiation is used to:
  (a) detect aircraft;
  (b) treat skin disorders;
  (c) study crystal structure;
  (d) treat cancerous growths;
  (e) give a suntan?

23 Static electricity

What effects are caused by static electricity?

1. A plastic comb rubbed against fabric or combed through your hair attracts hair, dust and small bits of paper.
2. The same charged comb will attract the water molecules in the stream of water from a tap.
3. A balloon rubbed against your sweater will stick to the wall or ceiling.
4. Cellulose acetate photographic film and other plastic objects such as records and lunch boxes tend to pick up dust, hair and fluff.
5. Clothes made of nylon and other synthetic fabrics tend to cling and crackle when you remove them. They become charged and cause small sparks. These can be seen in the dark.

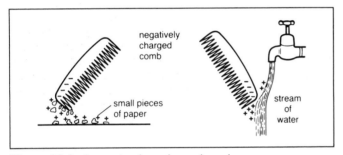

Figure 23.1 *Attraction by a charged comb:*

Positive charges are induced on the paper and water. These are then attracted to the comb.

What happens when charges are transferred?

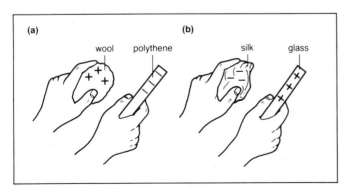

Figure 23.2 *Transferring charge by friction:*

When certain pairs of materials are rubbed together they become charged. Examples are polythene and woollen fabrics (a), or glass and silk (b).

● Electrons have a negative charge. They are transferred by the friction of the rubbing process from one material to the other.
● Some materials *gain* extra electrons. They become **negatively** charged. Examples are polythene and silk.
● Some materials *lose* electrons. They become **positively** charged. Examples are glass, wool, perspex and cellulose acetate.

When polythene and wool are rubbed and then separated, *equal* and *opposite* charges are left on the two materials.

How are insulators and conductors different?

Insulators

The electrons of the atoms in an insulator are not free to move around. Extra charges may be given to the surface of an insulator. But they are unable to move through it or across its surface. When given a charge, insulators stay charged unless surface moisture or dust let it leak away.

Some **insulators** are: glass, rubber, most plastics and wax.

Conductors

These have vast numbers of very mobile electrons. The electrons transport electric charge all over the material. Charges spread out to an even level (voltage or potential) on the surface of a conductor. Conductors should be insulated from the ground, the bench or your hands. Otherwise any extra charge on a conductor will be carried away by the electrons and leak to the Earth.

Some **conductors** are: all metals, water-containing solutions and materials, graphite.

What causes attraction and repulsion?

● Only two types of charge exist.
● *Like* charges always *repel* each other.
● *Unlike* charges always *attract* each other.

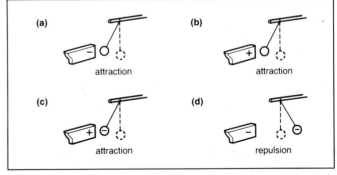

Figure 23.3 *Only repulsion confirms that an object is charged:*

An *uncharged* ball is attracted to both a negatively charged strip (a) and a positively charged strip (b). So attraction (c) cannot be taken as proof that an object has (the opposite) charge. It might equally be uncharged. But when *repulsion* occurs as in (d) we can be sure that the ball is charged. Also, we can say it must have the *same* charge as the strip.

How are charges induced?

The uncharged ball in figure 23.3 is attracted to the charged strips (a) and (b). This is because it gains an **induced** charge.

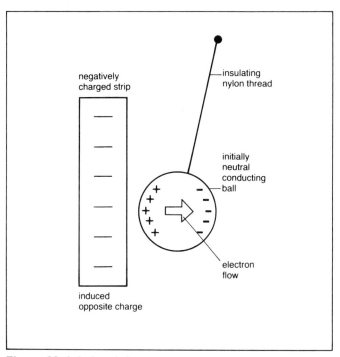

Figure 23.4 *Induced charges:*

Like charges repel each other. So the electrons in the neutral ball move to its far side. This leaves a shortage of electrons on the near side. The *induced* positive charge results from the shortage of electrons on the near side of the ball is responsible for its attraction to the strip. If the ball is touched with a finger this will allow some of the electrons on the ball to flow further away to ground. They will take their negative charges with them. If the finger is then removed before the strip, the ball will keep a net positive charge. This is because of the missing electrons. This is called **charging by induction**.

● The total charge on the ball remains zero, unless it is touched or connected to the ground.
● Induced charges are always *opposite* charges. They therefore cause attraction.
● The same induction process explains how the charged comb attracts small bits of paper. Also it shows why dust and fluff are attracted to charged plastic objects.

Applications

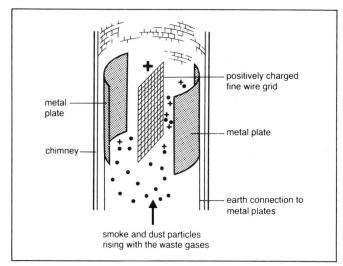

Figure 23.5 *Dust precipitation:*

An electrostatic precipitator removes smoke and dust particles from the waste gases going up the chimneys of factories and power stations. The high voltage on the wire grid causes a stream of positive ions to flow from the grid towards the metal plates. These ions attach themselves to the dust particles. These are then attracted to the earthed metal plates. From time to time the plates are tapped. So the attached smoke and dust falls down the chimney to be removed at the bottom.

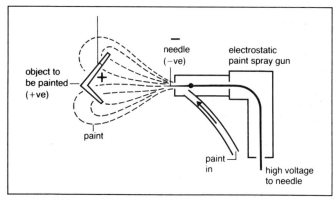

Figure 23.6 *Paint spraying with an electrostatic spray gun:*

The fine needle at the tip of an electrostatic paint spray gun is charged negatively to about 75 kV. It gives all the small droplets of paint in the spray a negative charge. The object to be painted is made positively charged. It then attracts the negatively charged paint droplets. This has two important effects:
● Less paint is wasted. This is because it is attracted to the object rather than missing it when passing through holes and past edges.
● The paint is more evenly spread over the object. It even reaches round corners.

24
Electric circuits

What is an electric current?

An electric current is caused by the movement of any charged particles through a conductor.

Examples are:
- electrons in metallic conductors (wires);
- electrons and ions in gases and liquids;
- electrons and holes in semiconductors.

The electric current in a conductor is a measure of the rate of flow of electric charge through it.

$$\text{current} = \frac{\text{charge}}{\text{time}}$$

$$I = \frac{Q}{t}$$

$$\boxed{\frac{Q}{I \times t}}$$

Charge Q is measured in **coulombs (C)**. **Time t** is measured in **seconds (s)**. **Current I** is measured in **amperes (A)**.
- A current of 1 ampere is a rate of flow of electric charge of 1 coulomb per second.

Example: A charge of 180 coulombs flows through a lamp every 2 minutes. What is the electric current in the lamp?
$Q = 180 \,\text{C}$ and $t = 2$ minutes $= 2 \times 60 \,\text{s}$.

$$I = \frac{Q}{t} = \frac{180}{2 \times 60} = 1.5 \text{ amperes}$$

Example: A battery circulates charge round a circuit for 30 seconds. The current in the circuit is 5 amps. How much charge flows through the battery?

$$\text{charge} = \text{current} \times \text{time}$$
$$Q = I \times t = 5 \times 30 = 150 \text{ coulombs}$$

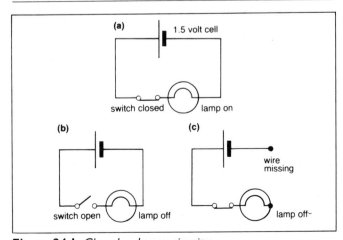

Figure 24.1 *Closed and open circuits:*

Currents need **closed** or complete circuits (a) in which to flow. A circuit may have a gap or break in it. This is called an **open** circuit (b and c). No current can flow in an open circuit. See page 157 for a table of circuit symbols.

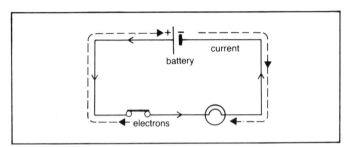

Figure 24.2 *Conventional current and electron flow:*

A **conventional current** is said to flow round a circuit from the positive terminal of the battery to its negative terminal. (This is shown by arrows on wires.) In metal wires the flow of **electrons** is in the *opposite* direction to the conventional current. Electrons flow from the negative terminal. They are attracted to the positive one. (This is shown by broken arrows).

Series connection

In **series circuits** the current passes through each component in turn. This is like a series of events, one after the other. If the filament in any one of the lamps breaks, then the current will stop everywhere in the open circuit.

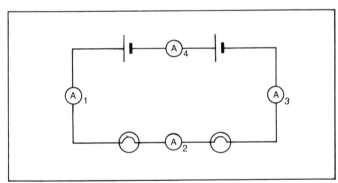

Figure 24.3 *Lamps and ammeters in a series circuit:*

Ammeters are used to measure electric currents.
- An ammeter must be connected *in series* with the circuit. This is so that the current flows through it.
- The current in a series circuit is the same at all points around the circuit.
- All four ammeters read the *same* current.
- The current does *not* get less as it goes round.

Parallel connection

Some circuits have branches. In these the current divides up. The branches are said to be connected **in parallel**. The divided currents flow side-by-side or in parallel.

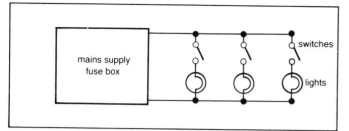

Figure 24.4 *Lights connected in parallel:*

Each one can be switched on independently. When *any one* switch is closed there is a complete circuit from the fuse box, through the switch and lamp, and back to the fuse box.

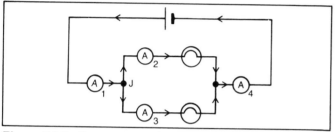

Figure 24.5 *Conservation of current:*

The *total* current must be the same at all points around a circuit. It follows that:

the sum of the currents in the branches of a parallel circuit = the current entering or leaving the parallel section

the readings of ammeter 2 + ammeter 3 = reading of ammeter 1 = reading of ammeter 4

At the junction J in the circuit the current entering the junction must equal the sum of the currents leaving the junction. For example, if that current 2 is 2 amps and current 3 is 3 amps, then currents 1 and 4 must both be 5 amps.

What is a potential difference or voltage?

Electric charge flows 'downhill' like water.

The electrical level or 'height' at a point in a circuit is called the voltage or the potential.

Using the conventional current direction, 'downhill' is towards the negative terminal of the battery. The battery is an **energy source** which does the job of pumping charge up to the 'top of the hill', so that it can then flow downhill as a current through the conductors in the circuit.
- The **potential difference** or **p.d.** between two points in a circuit is measured in **volts (V)**. So it is often called the 'voltage'.
- **Voltmeters** are used to measure voltages.

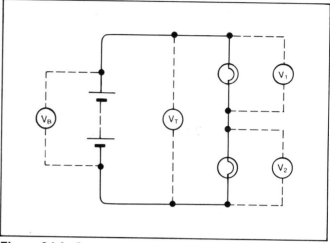

Figure 24.6 *Connecting a voltmeter to a series circuit:*

When a voltmeter is used to measure the voltage across a conductor in a circuit, it should be connected *in parallel* with it.

The voltmeter readings show that $V_1 + V_2 = V_T$. Also $V_T = V_B$, the voltage across the battery terminals.

the sum of the voltages in a series circuit = the voltage across the battery or energy source

How much energy is supplied by a battery?

The battery uses stored energy to move electric charge round a circuit through the conductors.

The p.d. or voltage across a conductor is the work done in joules when 1 coulomb of charge is passed through it. (1 volt = 1 joule per coulomb).

$$\text{p.d. (voltage)} = \frac{\text{work done}}{\text{charge moved}}$$

$$V = \frac{W}{Q}$$

Example: An electric heater produces 8.64 kJ of heat when 36 C of charge are passed through it. What is the voltage across it?
The work done by power supply = 8.64×10^3 J.

$$V = \frac{W}{Q} = \frac{8.64 \times 10^3}{36} = 240 \text{ volts}$$

Example: A 12 volt car battery passes 800 coulombs of charge through a starter motor. How much energy is supplied?

energy supplied = work done passing the charge
$$W = QV = 800 \times 12 = 9600 \text{ joules}$$

Resistance

Resistance arises in all components of a circuit. The resistance is where the electrons give up the potential energy they carry from the battery or power source.

For example, the resistance of a lamp filament causes the energy carried by the electrons to be converted into heat and light energy.

The resistance of a conductor depends on:
● its length;
● its cross-sectional area;
● the material of which it is made.

The resistance of a wire increases in proportion with its length. Doubling the length doubles its resistance.

The resistance of a wire depends inversely on its cross-sectional area. Doubling the diameter or thickness quarters its resistance.

How do we measure resistance?

The resistance of a conductor tends to *oppose* the current in a circuit. A potential difference or voltage across a conductor is needed for current to flow through it.

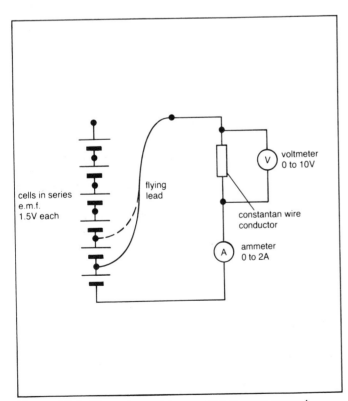

Figure 25.1 *Measuring resistance using an ammeter and voltmeter.*

The wire conductor is connected to 1, 2, 3 and more cells in turn. A set of readings for current and voltage can be then obtained as below:

V in volts	0	1.5	3.0	4.5	6.0	7.5	9.0
I in amps	0	0.25	0.5	0.75	1.0	1.25	1.5
R in ohms		6	6	6	6	6	6

The ratio V/I remains constant for this wire. This is called its *resistance*.

$$\text{resistance} = \frac{\text{voltage}}{\text{current}}$$

$$R = \frac{V}{I}$$

● Resistance is measured in **ohms** (Ω).
● The ohm is the resistance of a conductor through which the current is 1 ampere when the p.d. between its ends is 1 volt.

Example: A current of 4 A flows through a car headlamp when it is connected to a 12 V car battery. What is its resistance?

$$R = \frac{V}{I} = \frac{12}{4} = 3\,\text{ohms}$$

Example: What voltage would be needed to drive a current of 0.2 A through a torch lamp of resistance 22.5 Ω?

$$V = IR = 0.2 \times 22.5 = 4.5\,\text{volts}$$

Ohm's law

The resistance of most conductors varies with temperature. George Ohm measured the resistance of metal wires. He found that the ratio V/I remained constant if the temperature was kept constant. His law states:

The current through a metallic conductor, if maintained at constant temperature, is directly proportional to the potential difference between its ends:
$$I \propto V$$

● Conductors which obey Ohm's law, (i.e. have a constant resistance) are called **ohmic conductors**.

Characteristics of conductors

A graph of the current *I* through a conductor against the voltage *V* between its ends, gives a curve known as the characteristic of the conductor.

The $-V$ and $-I$ values show what happens when the connections to the battery are reversed and current flows in the opposite direction.

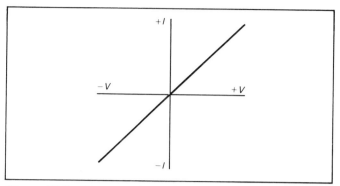

Figure 25.2 *Ohmic conductor (e.g. constantan wire):*

This is a straight-line characteristic. It passes through the origin of the graph. The wire has a *constant* resistance which is the same for both directions of current flow. It obeys Ohm's law.

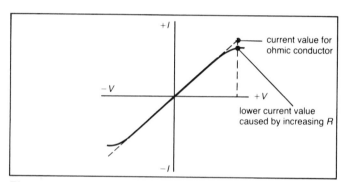

Figure 25.3 *Filament lamp:*

At low currents the characteristic may be fairly straight. However, as the current rises more heat is produced. The temperature rise increases its resistance, so at a particular voltage where the filament temperature has risen (dotted line), the current value is lower than it would be for an ohmic conductor.

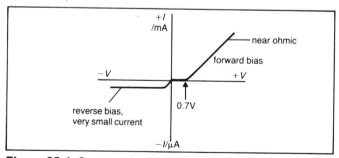

Figure 25.4 *Semi-conductor diode (silicon):*

This device conducts well in one direction, called **forward bias**. But it passes almost no current in the other direction, called **reverse bias.**

In the forward bias direction its resistance is quite low. Apart from needing about 0.7 volts before it starts conducting, it is nearly ohmic. In the reverse bias direction its resistance is very high. It is typically millions of ohms.

Resistors connected in series

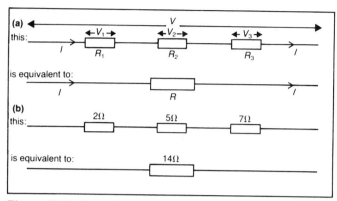

Figure 25.5 *Equivalent resistors (series):*

The current flowing in a circuit always depends on the *total* resistance in the circuit. When conductors or resistors are connected together in series, because the current must flow through all of them, their resistances add together.

● Resistors in *series* carry the same *current*. In figure 25.5 the total resistance R is given by the equation for *series* resistors:

$$R = R_1 + R_2 + R_3$$

Resistors connected in parallel

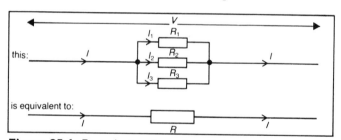

Figure 25.6 *Equivalent resistors (parallel):*

Resistors connected in parallel have a smaller combined resistance. They conduct a larger total current than they do separately. This is because the parallel paths make it easier for the current to flow.

● Resistors in *parallel* have the same voltage across them. In figure 25.6 the total resistance R is given by the equation for *parallel* resistors:

$$\frac{1}{R} = \frac{1}{R_1} + \frac{1}{R_2} + \frac{1}{R_3}$$

Example: A $6\,\Omega$ resistor is connected in parallel with a $3\,\Omega$ resistor. What is their combined resistance?

$$\frac{1}{R} = \frac{1}{R_1} + \frac{1}{R_2} = \frac{1}{6} + \frac{1}{3} = \frac{1+2}{6} = \frac{3}{6} = \frac{1}{2}$$

$$\therefore R = 2 \text{ ohms}$$

Review questions: Chapters 23, 24 and 25

C23 1 Copy and complete the following:
   (a) A material which _____ extra electrons becomes negatively charged.
   (b) Materials which lose electrons become _____ charged.
   (c) When materials become charged by rubbing they have _____ and opposite charges.
   (d) Only _____ confirms that an object is charged.

2 Which of the following are examples of the effects of static electricity:
   (a) a balloon sticking to the wall;
   (b) a car door giving a shock in dry weather;
   (c) a shock received from a live mains cable;
   (d) a nylon jumper crackling when taken off?

3 (a) Are electrons in insulators free to move?
   (b) Are the electrons mobile in conductors?
   (c) How many types of charge exist?
   (d) Do like charges attract or repel each other?

4 Which of the following are insulators:
   (a) glass; (b) copper; (c) graphite; (d) wax;
   (e) perspex; (f) rubber (g) aluminium?

5 Describe the effect of hanging a negatively charged rod near another hanging rod which is:
   (a) negatively charged;
   (b) positively charged;
   (c) uncharged.

6 The figure shows an initially uncharged conducting ball with a negatively charged strip nearby.

   (a) Copy the diagram and show the induced charges on the ball.
   (b) Describe the effect of touching the ball with a finger before the strip is removed.
   (c) What is the final charge on the ball?

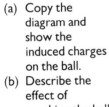

insulating nylon thread

7 (a) Dust precipitators use a high voltage to produce a stream of ions. What do the ions become attached to? Where are they then attracted?
   (b) How are the paint droplets charged in an electrostatic paint spray gun?

C24 8 What letters are used to represent:
   (a) current; (b) charge; (c) voltage; (d) ampere;
   (e) coulomb; (f) volt; (g) energy or work?

9 What is the current flowing in:
   (a) a light bulb carrying 120 coulombs every 4 minutes;
   (b) a cooker carrying 54 000 coulombs every 30 minutes?

10 Find how much charge passes through:
   (a) a soldering iron carrying 3 A in 10 min;
   (b) a television set carrying 0.8 A in 1 h.

11 The figure shows an electric circuit.
   (a) What do the three symbols in the diagram represent?
   (b) Is this circuit shown open or closed?

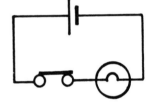

   (c) Copy the circuit. Indicate the direction of flow of conventional current.
   (d) Which way will electrons flow?

12 (a) Draw a circuit with two cells and two lamps in series. Add four ammeters connected in different positions in the circuit.
   (b) If one ammeter reads 0.2 A, what will the other ammeters read?

13 Copy and complete the following:
   (a) An electric current is the flow of _____ particles in a conductor.
   (b) A current is the rate of flow of _____.
   (c) Current is measured with an _____ connected in _____ with a circuit.
   (d) The currents into a junction _____ the currents out of the junction.

14 What are the SI units of:
   (a) current; (b) charge; (c) voltage; (d) energy?

15 Copy and complete the following:
   (a) Voltmeters measure _____.
   (b) A voltmeter is connected in _____ across a component.
   (c) The sum of the voltages in a series circuit equals the voltage across the _____.

16   I volt is I joule per coulomb. What are:
(a) 2 volts;   (b) 5 volts;   (c) 240 volts?

17   What is the voltage across:
(a) an electric heater if 2.4 kJ of heat are produced when 10 C of charge pass through it?
(b) a lamp if 36 J of heat and light are produced when 3 C of charge flow?

18   How much energy is supplied by a:
(a) 12 V battery when it passes 6 C of charge through a lamp;
(b) 5 V battery when is passes 1 C of charge through a microcomputer?

C25   19   (a) What does resistance in a circuit oppose?
(b) If a length of wire is doubled, how does its resistance change?
(c) What is the SI unit of resistance?
(d) What ratio does the resistance equal?

20   What size battery is needed to drive a current of:
(a) 2 A through a resistor of 4 Ω;
(b) 0.06 A through a lamp of resistance 100 Ω?

21   Calculate the current flowing in:
(a) a lamp of resistance 100 Ω, when connected to a 5 V power supply;
(b) an electric kettle of resistance 30 Ω? (The mains voltage is 240 V.)

22   What is the resistance of:
(a) a 12 V heater with a current of 3 A;
(b) a resistor with a current of 0.1 A when connected to a 10 V power supply?

23   Copy and complete the following:
(a) _____ law says that the current through a metallic conductor, kept at a constant _____, is _____ to the voltage.
(b) Conductors which obey this law are called _____.

24   The figure shows a current–voltage graph.
(a) What is this type of graph called?
(b) Is this graph for a diode, a resistor or a filament lamp?

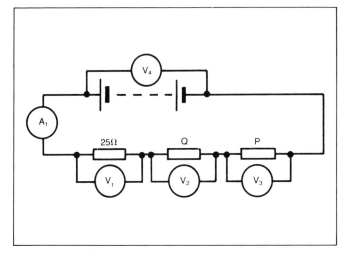

(c) What happens to the current for equal increases in voltage?
(d) What happens to the resistance of the device as the voltage increases?

25   (a) What device conducts in one direction only?
(b) What is this direction called?

26   What is the total resistance of:
(a) two resistors of 100 Ω in series;
(b) a 500 Ω and a 300 Ω resistor in series;
(c) two 20 Ω resistors in parallel;
(d) a 4 Ω and a 6 Ω resistor in parallel?

27   The figure shows combinations of resistors.

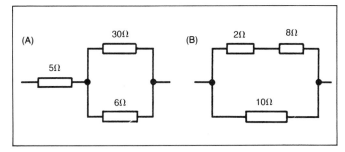

For A:
(a) Which resistors are in parallel?
(b) What is the combined resistance of the 6 Ω and the 30 Ω resistors?
(c) What is the total resistance in A?
For B:
(d) Which resistors are in series?
(e) What is the combined resistance of the 2 Ω and the 8 Ω resistors?
(f) What is the total resistance in B?

28   The figure shows a series circuit.

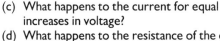

(a) If V_1 reads 5 V, what is the reading of A_1?
(b) If Q = 10 Ω, what is the reading of V_2?
(c) If V_4 reads 10 V, what is the value of P?

26 Magnetism

Which materials are magnetic?

Some materials like iron are strongly affected by magnetism. They are called ferromagnetic. A magnet will attract anything containing iron, from a lump of iron ore to a steel ball-bearing or a safety pin. The only other metals strongly attracted by magnets are cobalt and nickel.

	Soft magnetic materials	Hard magnetic materials
Examples	Soft iron	Steel, special alloys of iron containing aluminium, nickel and cobalt
Names	Stalloy, mumetal	Alnico, ticonal, magnadur
Can be magnetised	Very easily	Less easily
Induced magnetism	Is temporary, can be switched on and off rapidly	Is permanent
Is used for making	Cores of electromagnets, transformers and electric motors	Permanent magnets e.g. loudspeakers, door catches and magnetic drain plugs on car engines

- All magnets have two **poles** called *north* and *south*.

How do we test for magnetisation?

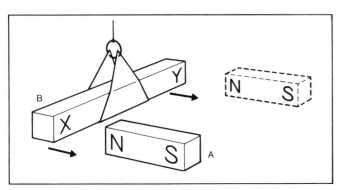

Figure 26.1 *Testing an iron bar:*

- Like poles repel.
- Unlike poles attract.
- Unmagnetised iron is also attracted to magnets.
- Repulsion is needed to prove magnetisation.

Test results.

Pole of magnet A	Action of suspended bar B	Conclusion: X is a
N	repelled	N pole
S	repelled	S pole
N	attracted	S pole *or* unmagnetised iron if end Y is also attracted

How can we show magnetic field patterns?

Tap a cardboard sheet sprinkled with iron filings. This allows them to move and turn. They will line up with a magnetic field. Each little filing becomes a small magnet. They join up in chains along the **magnetic field lines**.

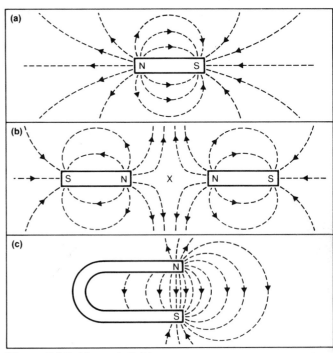

Figure 26.2 *Magnetic field patterns:*

The dotted lines show the *shape* of the magnetic field. A plotting compass also shows the *direction* of the magnetic field lines.
  (a) A bar magnet: the lines go from north to south.
  (b) Repelling magnets: X is a neutral point between them.
  (c) A U-shaped magnet has a strong magnetic field between its poles.

What is induced magnetism?

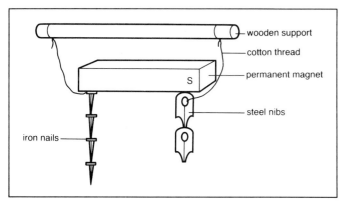

Figure 26.3 *Induced magnetism:*

The nails and nibs are not magnetised at first. They are given **induced magnetism** by the permanent magnet. Then they are attracted to each other. When the magnet is removed:
- The iron nails all quickly drop off. This shows that they had **temporary** induced magnetism.
- The steel nibs continue hanging in a chain. This shows that their induced magnetism has become **permanent.**

How can we make a magnet?

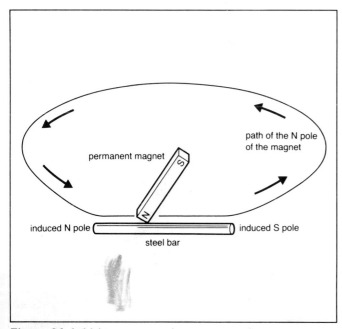

Figure 26.4 *Making a magnet by magnetic induction:*

Stroke the bar with one pole of a permanent magnet. This lines up all the tiny imaginary magnets (known as magnetic dipoles) inside the bar. Opposite poles attract. So the north pole of the permanent magnet leaves an induced south pole at the end of the bar.

How should we store magnets?

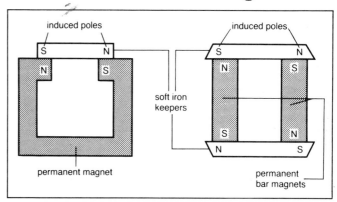

Figure 26.5 *Soft iron keepers help magnets stay strongly magnetised while they are stored:*

Magnetism is induced in the keepers. This forms a closed loop of magnetic material which is very strong and resists demagnetisation.

The Earth's magnetic field

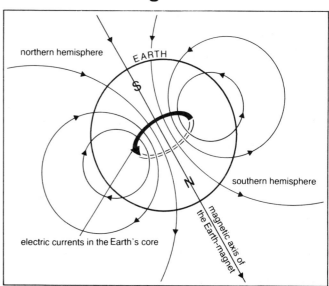

Figure 26.6 *The Earth's magnetic field can be explained by large electric currents in the core:*

In effect, the Earth is a large electromagnet. It has a *south* magnetic pole in the *northern* hemisphere some 1000 km from the true North Pole.

The angle of declination is the angle on the Earth's surface between the direction of the true North Pole and the magnetic pole.

In the northern hemisphere the Earth's magnetic field goes down into the ground. This can be seen from the map of the Earth's field lines above. A dip circle is used to measure the **angle of dip**. This is the angle by which the Earth's magnetic field dips below the horizontal.

27 Effects of electricity

What are the magnetic effects of a current?

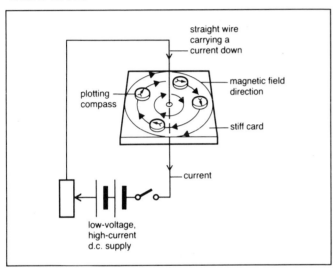

Figure 27.1 *The magnetic field around a wire:*

The magnetic field goes in circles around a wire carrying a current. Maxwell's screw rule may be used to predict the direction of the magnetic field.

Maxwell's screw rule

Turn a right-handed screw so that it moves forwards in the same direction as an electric current. It is then turning in the direction of the magnetic field.

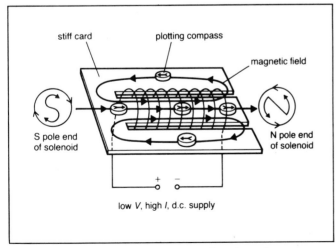

Figure 27.2 *The magnetic field around a solenoid:*

The solenoid has a magnetic field which is similar to that of a bar magnet.

A bar of soft iron can be fitted inside a solenoid as a core. This turns it into a powerful **electromagnet**. When the current in the solenoid is switched off, the temporary magnetism in the soft iron core will quickly disappear. So the electromagnet can be switched on and off. This makes it useful for lifting iron and steel components in industry.

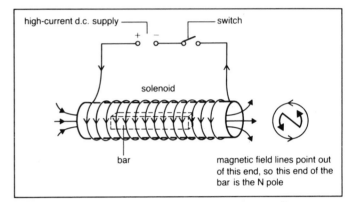

Figure 27.3 *Making a permanent magnet electrically:*

A high *direct* current is needed only for a short time to produce the strongest permanent magnetisation. You need a solenoid with several hundred turns of insulated wire. A bar of steel or magnetically 'hard' material is placed inside it to be magnetised.

Demagnetising a steel object

To demagnetise a steel object, *slowly* pull it out of the magnetic field of a solenoid. An *alternating* current in the solenoid reverses the magnetisation of the object. The strength of the magnetisation gradually climbs down step by step every time the magnetic field reverses while the object moves further out of the field.

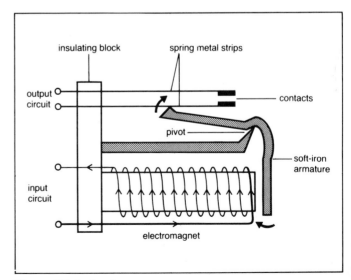

Figure 27.4 *A relay:*

This is a switching device which uses an electromagnet. Only a small current is needed at the input to operate the electromagnet. The magnet attracts a soft-iron armature which opens or closes a pair of contacts (a switch) in the output circuit. The contacts are normally open (as drawn). The output circuit can control (switch on or off) another circuit which perhaps carries a high current. There is no *electrical* connection between the input and output circuits.

What is the heating effect of a current?

Experiments show that the amount of heat produced by an electric current depends on:
- the time t (measured in seconds);
- the resistance R of the conductor;
- the square of the current: I^2.

The electrical energy W converted or heat energy produced can be calculated. Use the equation:

$$W = I^2 R t \quad \text{or} \quad W = I t V \text{ (joules)}$$

Example: A current of 4.0 A is passed through a thin cable for 1.0 h. Its resistance is 20 Ω. How much electrical energy will be converted to heat in the cable?

$$t = 1.0 \times 60 \times 60 \,\text{s} = 3600 \,\text{s}$$

Electrical energy converted $W = I^2 R t$:
$$\therefore W = (4.0)^2 \times 20 \times 3600 = 1.15 \times 10^6 \text{ joules}$$

What is electrical power?

Power is the rate of energy conversion.

$$\text{power} = \frac{\text{energy converted}}{\text{time}}$$

$$P = \frac{W}{t}$$

- The SI unit of power is the **watt (W).**
- 1 watt = 1 joule per second.

Example: A microwave oven converts electrical energy into microwave energy. It is used for 2 minutes to heat some food. The current is 5.0 amperes. The voltage is 240 volts. What is the electrical energy converted? What is the power of the oven?

$$t = 2 \text{ minutes} = 2 \times 60 \text{ seconds} = 120 \,\text{s}$$

$$W = I t V = 5.0 \times 120 \times 240 = 144\,000 \text{ joules}$$

So electrical energy converted = 144 kJ.

$$\text{power} = \frac{W}{t} = \frac{144 \,\text{kJ}}{120 \,\text{s}} = 1.2 \text{ kilowatts}$$

How much electrical energy will be needed to run a 200 watt electric motor for 5 minutes?

$$t = 5 \text{ minutes} = 5 \times 60 \text{ seconds} = 300 \,\text{s}$$
$$\text{energy } W = P t = 200 \times 300 = 60\,000 \text{ joules}$$

Power in circuits

From: energy $W = I^2 R t$ and $W = I t V$

we get: power $P = \dfrac{W}{t} = \dfrac{I^2 R t}{t}$ and $P = \dfrac{I t V}{t}$

$$\therefore \text{ power } P = I^2 R \quad \text{and} \quad P = IV$$

- The fomula $P = IV$ is often remembered in terms of its units:

$$\text{watts} = \text{amps} \times \text{volts}$$

Example: An electric kettle takes a current of 12.5 amperes from the 240 volt mains. What is the power?

$$\text{power } P = IV = 12.5 \times 240 = 3000 \text{ watts or } 3.0 \,\text{kW}$$

Example: A 60 W electric light is used on the 240 V mains supply. What current will flow in the lamp?

$$\text{current } I = \frac{P}{V} = \frac{60}{240} = 0.25 \text{ amperes}$$

Fuses

To 'fuse' means to melt. An **electrical fuse** is a short thin piece of wire with a fairly low melting point. When the current through it gets too large, it gets hot. It then melts or 'blows'. This stops the current like a switch. Fuses are needed to protect against the fire risk caused by the heating effect of an electric current. When the current in a cable or appliance gets too large due to a fault, the fuse will break the circuit.
- The correct **rating** or fuse value to fit in a plug is the standard value *just above* the normal current required by the appliance.
- 1A, 2A, 3A, 5A and 13A are common fuse values.

Example: The power rating given on an electric iron is 960 W. The mains is 240 V. What fuse should be fitted?

$$\text{The normal 'safe' current: } I = \frac{P}{V} = \frac{960}{240} = 4.0 \,\text{A}.$$

The standard fuse value just above 4 A is 5 A.

28
Electricity at home

How do we use electricity at home?

Using electricity.

Heating	Lighting	Electric motors	Electronic equipment
Hair drier Electric kettle Immersion heater Electric cooker Electric fires Toaster	Filament lamps Fluorescent lamps Warning lights	Sewing machines Power tools Washing machines Vacuum cleaners Hair driers Food mixers Pumps in fridges and freezers	TV Radio Hi-fi Computers Microwave cookers

- The appliances in the first column are designed to convert electrical energy into heat energy. All of these appliances use a lot of energy. They are expensive to run. Many electrical machines in the home produce waste heat. For example, feel how hot a food mixer gets. Feel the warm air blown out of a vacuum cleaner.

What are the hazards of mains electricity?

- **Electric shock** can result from touching a live wire. This can be fatal.
  People are more at risk when:
  (i) Using portable tools such as hedge-trimmers, electric drills and electric irons. The cable can become damaged and expose live wires.
  (ii) In a damp environment such as a bathroom or washroom and when standing on a stone or solid tiled floor (in bare feet). Water on your hands or feet can allow a larger current to flow through you to the ground.
  (iii) They have a heart illness or weakness.
- **Fire** may be caused by faulty appliances, perished insulation, short-circuits and overloaded cables and sockets.

How do we make electricity safe?

The earth wire

- The earth wire is coloured green with a yellow stripe. It must be connected to the metal case of an appliance.

- Earth wires must have a good low-resistance connection to earth. This is in case a fault occurs and a current flows through the live and earth wire in series. The fuse in the live wire will blow. This will cut off the supply.

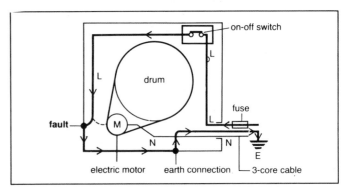

Figure 28.1 *A washing machine protected by an earth wire:*

There is a fault from the loose live wire L touching the case. But the circuit is completed through the machine case and the earth wire to the ground. (Follow the arrows.) This blows the fuse and protects us. (L = live wire. N = neutral wire. E = earth wire.)

Double insulation

Some appliances are **double insulated**. They are marked with the symbol ▣ . Examples are vacuum cleaners, hair driers and food mixers.

- The electric cable is insulated from the internal metal parts of the machine.
- All the internal metal parts are enclosed in an insulating plastic case. No external screws, handles or attachments make any direct connection with any internal metal part. For example the metal whisks of a food mixer will be mounted in plastic insulating bushes.
- The risk of receiving an electric shock is greatly reduced. So these appliances will not have an earth wire.

Fuses and circuit breakers

- **The consumer unit or 'fuse box'**. This is fitted with a fuse or **miniature circuit breaker (mcb)**. This protects each circuit from overload. The mcb has a switch which trips when the current exceeds a certain value. It is reset after the fault has been repaired.
- **An earth leakage circuit breaker (elcb) or residual current device (rcd)**. This is an extra circuit-breaking switch. It can be fitted to high-risk machinery such as electric lawnmowers, or to whole houses. These devices can detect a small current difference between the live and neutral wires. After a few milliseconds they then switch off the electricity. The speed with which these devices act saves people from a serious electric shock.
- **The fused plug.** A **cartridge fuse** is fitted in a plug to protect an appliance and its cable from the heating effect of a large current. Such a current might flow if a fault such as a short-circuit developed.

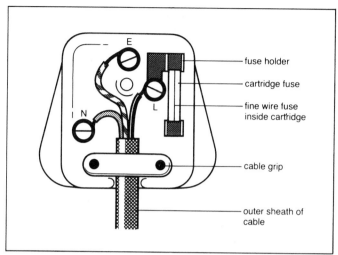

Figure 28.2 *Wiring a 13A fused plug.* (E = earth wire (green/yellow). L = live wire (brown). N = neutral wire (blue).)

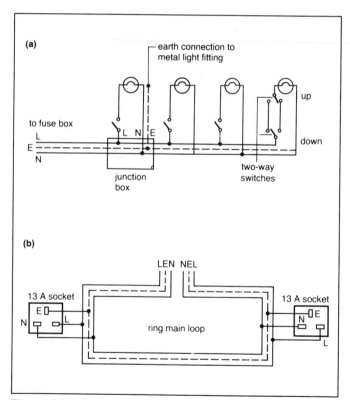

Figure 28.3 *House lighting and power socket ring-main circuits:*

(a) The lamps are connected in parallel with spurs of wire radiating out of junction boxes. There is a plastic junction box for each light. There are two-way switches for the top and bottom of the staircase. These circuits are called '**radial**'.

(b) The sockets on a '**ring-main**' circuit are also connected in parallel. The use of a 'ring' of wire reduces the gauge or thickness of wire which has to be used. Both ends of the loop are connected to the fuse box.

What is the cost of electricity?

The kilowatt hour unit

We pay for the *energy* used or converted in our homes: energy converted = power × time.
- The unit of energy for which we pay is the **kilowatt hour (kWh)**.
- 1 kilowatt hour = 1000 watts × 1 hour
  $\therefore$ 1 kWh = 1000 watts × 3600 seconds
  $= 3.6 \times 10^6$ joules = 3.6 MJ
- The number of 'units' used or converted is given by:

$$\text{number of kWh units} = \text{number of kilowatts} \times \text{number of hours}$$

$$\text{cost} = \text{number of kWh units} \times \text{price per kWh unit}$$

Example: How many kWh units of electrical energy will be used in a day by: (a) a 3 kW electric fire and (b) a 60 W electric lamp? What will each cost if the price of a unit of electricity is 5p?
The number of units = kilowatts × hours:
(a) The fire will use: $3\,kW \times 24\,h = 72\,kWh$

(b) The lamp will use: $\dfrac{60\,W}{1000} \times 24\,h = 1.44\,kWh$

The cost = kWh units × price per unit:
(a) The fire will cost: $72 \times 5p = 360p$ or £3.60
(b) The lamp will cost: $1.44 \times 5p = 7.2p$

How do we read an electricity meter?

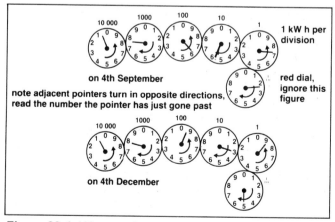

Figure 28.4 *When reading a dial (or digital) meter, ignore the last figure (tenths of units).*

Reading on December 4th is 07928.
Reading on September 4th is 07657.

$\therefore$ number of units used = 271 kWh

29 Motors and dynamos

What is the motor effect of a current?

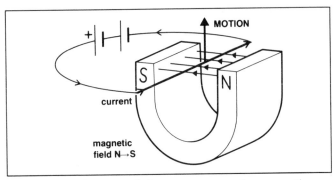

Figure 29.1 *The catapult force on a current in a magnetic field:*

The wire is catapulted out of the magnetic field of the magnets.
- The **catapult force** or **thrust** acts at 90° to both the current and the magnetic field.
- When either the current or the magnetic field direction is reversed the direction of the movement of the wire is reversed.

Fleming's left-hand rule, the motor rule

Hold the thumb and first two fingers of your left hand at right angles to each other. Point the:
- **f**irst finger in the **f**ield direction (N to S);
- se**c**ond finger in the **c**urrent direction;
- the **th**umb now points with the **th**rust.
  Motion is in the direction of the thrust.

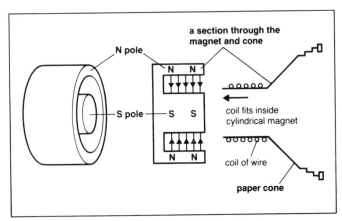

Figure 29.2 *The moving-coil loudspeaker:*

When an alternating current flows through the coils the paper cone moves in and out of the magnetic field. This produces sound waves in the air.

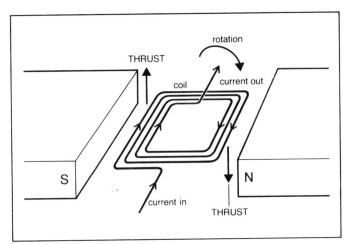

Figure 29.3 *Turning a coil:*

A coil of wire with a d.c. current between the poles of a magnet will rotate. The direction can be found using Fleming's left-hand rule on the coil sides. The turning effect or moment of the forces acting on the coil can be made stronger by:
- increasing the current in the coil;
- putting more turns on the coil;
- using a stronger magnetic field.

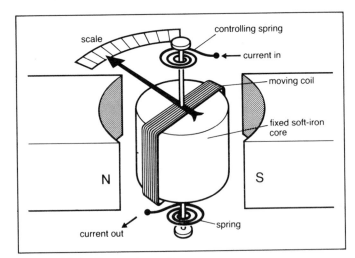

Figure 29.4 *The moving-coil meter:*

The moving-coil meter can only measure *direct* currents. It can be made more sensitive (i.e. detect smaller currents) by:
- using a finer or weaker controlling spring;
- using a stronger magnet;
- putting more turns on the coil;
- making the coil have a larger area.

Using a radial magnetic field gives the meter a scale with evenly spaced divisions.

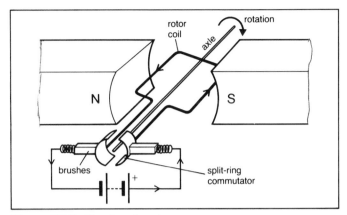

Figure 29.5 *The d.c. electric motor:*

- The coil which rotates is called the **rotor**.
- The device which supplies current to the rotor is called a **commutator.**
- To keep a coil turning it is necessary to keep reversing the current. In d.c. motors the **split-ring commutator** acts as an automatic current-reversing switch. The two halves of the ring are connected to the two ends of the rotor coil. The split-ring is mounted on the axle and turns with the coil. Each time another half of the ring meets a brush the current flows the opposite way round the coil.

Commercial electric motors.

Design improvements	Purpose or effect
Replace magnets with field coils (stators)	More powerful motor, runs on a.c. and d.c.
Wind rotor on soft-iron core	Magnetic field stronger, more power for same I
Use rotor with several coils	Runs more smoothly, more power

What is the dynamo effect?

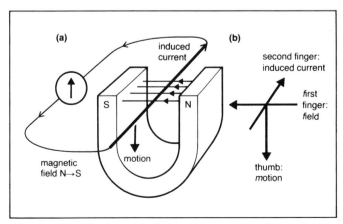

Figure 29.6 *The dynamo effect and Fleming's right-hand rule:*

(a) When a wire cuts a magnetic field a p.d. or voltage is induced between the ends of the wire. This voltage is called an electromotive force or e.m.f. The size of the induced e.m.f. depends on:

- how fast the wire cuts the magnetic field;
- the strength of the magnet.

In a *closed* circuit a current is induced.

(b) The *direction* of the induced current (second finger) depends on:
- the direction of the **m**otion of the wire (thumb);
- the direction of the magnetic **f**ield (first finger).

This is **Fleming's *right*-hand dynamo rule.**

Currents will also be induced if a magnet is moved into or out of a coil of wire. The *size* of the induced current depends on:
- the speed with which the magnet is moved into or out of the coil (when the magnet is held still there is no induced current at all);
- the number of turns on the coil.

The *direction* of the induced current reverses when the motion of the magnet is reversed. It also reverses when the poles are reversed.

Electromagnetic induction

Lenz's law:

The direction of an induced current is such as to oppose the change or motion causing it.

Faraday's law of electromagnetic induction:

The size of the induced e.m.f. depends directly on how quickly a wire cuts a magnetic field. Alternatively it may depend on how quickly the strength of a magnetic field changes.

Applications of electromagnetic induction

The e.m.f. induced in a coil or wire can be used to drive a current round a circuit.

1. The rotor of an electric motor can be spun. It then will work as a generator or dynamo.
2. A magnetised strip of metal tape can be moved past the poles of an electromagnet. An e.m.f. is then induced in the coils. A tape-recorder play-back head uses this effect.
3. Sound can make a coil vibrate between the poles of a magnet. We then have a moving-coil microphone generating an e.m.f.
4. In a transformer the changing current in one coil makes a magnetic field change inside another coil. This induces an e.m.f. in the second coil.

Review questions: Chapters 26, 27, 28 and 29

C26

1 (a) What are the names of the poles of a magnet?
(b) Do like or unlike poles attract?
(c) Would an electromagnet be made from soft or hard magnetic materials?
(d) Is *induced* magnetism in hard materials permanent?
(e) What test is needed to prove that an object is magnetised?

2 (a) Draw a diagram to show the imaginary field lines around a bar magnet.
(b) The lines go from where to where?
(c) Draw a diagram to show the field lines around two identical magnets with their south poles facing each other.
(d) What is true about the magnetic field midway between the two like poles?

3 The figure shows a permanent magnet attracting iron nails and steel nibs.

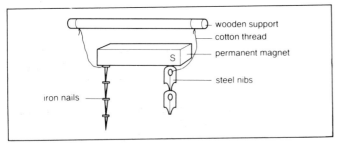

wooden support
cotton thread
permanent magnet
S
steel nibs
iron nails

(a) When the magnet is removed what will happen to: (i) the nails; (ii) the nibs?
(b) Which were temporarily magnetised?

4 Copy and complete the following:
(a) A magnet can be made by _____ a steel bar with a permanent magnet.
(b) A soft iron _____ is used to help keep a magnet strongly magnetised.

5 A bar magnet is suspended from a support.
(a) Which direction will the North Pole face?
(b) What does this mean about the Earth's magnetic poles?
(c) What is the name of the angle on the Earth's surface between the true North Pole and the magnetic pole?

C27

6 (a) What type and shape of field surrounds a wire carrying a current?
(b) Which rule is helpful for finding the direction of this field?
(c) What are the field lines like inside a solenoid?

7 The figure shows a vertical wire through a horizontal card. Copy the diagrams and add lines to show the direction of the magnetic fields due to the currents.

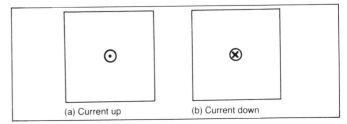

(a) Current up   (b) Current down

8 If a steel bar is inside a solenoid:
(a) What type of current is used to permanently magnetise the bar?
(b) While the bar is removed, what type of current can demagnetise the bar?

9 A student decides to make an electromagnet. She winds *insulated copper wire* around a U-shaped *soft iron* core. She then connects it to a *switch* and *battery*.
For the words in *italics*, explain the reasons for her actions.

10 How much electrical energy will be converted into heat in the following:
(a) a $50\,\Omega$ resistor carrying 1 A for 10 s;
(b) a cooker ring of resistance $100\,\Omega$, carrying 2 A for 5 min;
(c) a 250 V fire with a current of 4 A for 1 h.

11 Calculate the power of:
(a) a TV set which takes a current of 0.8 A from a 240 V mains supply;
(b) a torch bulb labelled 2.5 V, 0.3 A.

12 Copy and complete the following:
(a) A short thin piece of _____ with a fairly _____ melting point is called a _____.
(b) If 1 A, 2 A, 3 A, 5 A and 13 A fuses are available, which of them should be used as protection in questions 10 and 11?

C28

13 Which of the following are designed to convert electrical energy into heat:
(a) radio; (b) kettle; (c) food mixer; (d) toaster; (e) hair drier; (f) sewing machine?

14 Which of the list in question 13 contain an electric motor?

15 Copy and complete the following:
The hazards of mains electricity include:
(a) electric shock from touching a _____ wire;
(b) fire caused by _____ appliances, such as worn _____, or _____ sockets;
(c) earth wires, double insulation, fuses and circuit breakers are examples of _____ systems.

16 The figure shows a standard three-pin plug. David wishes to connect the plug to his new cassette recorder.
It has blue, brown and green/yellow wires.

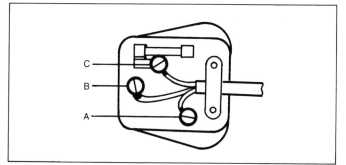

(a) What is the name of each coloured wire?
(b) Where should he connect each wire?
The plug is sold with a 13 A fuse in it.
(c) What size fuse should David put in the plug if the cassette recorder uses 1.5 A?

17 (a) Are lamps in a house wired in series or parallel?
(b) A 60 W lamp is run for 4 hours. How many kWh of energy are used?
(c) How much will it cost if electricity is 5p per unit?

18 Tessa wishes to check her electricity meter. Three months ago it read 5142. The figure shows the reading now.

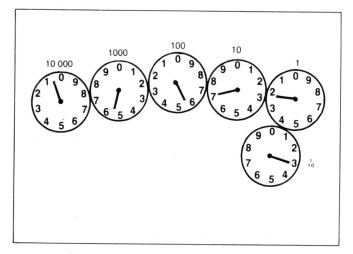

(a) What is the reading now?
(b) How many units has she used?

C29 **19** The figure shows a horizontal wire between the poles of a magnet. When the switch is closed:
(a) Which way will the current flow?
(b) What is the direction of the magnetic field?
(c) Which way will the wire move?

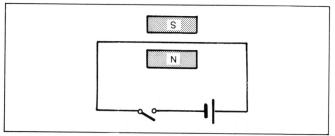

(d) What change could be made to the circuit to make the wire move the opposite way?

20 The figure shows a view looking down on to a horizontal coil of wire between the poles of a magnet. (Assume the field is uniform between the poles.) If the current flows as shown on the diagram:
(a) What direction is the thrust on side: (i) AB; (ii) CD?
(b) Describe which way the coil will rotate.
(c) What change could be made to make the turning effect of the forces bigger?

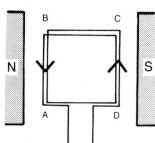

21 (a) What does a moving-coil meter measure?
(b) How can a moving coil-meter be made more sensitive without changing the coil?
(c) In a d.c. motor, what is the coil called?
(d) What is the device used to carry the current to the coil called?
(e) Why is this device a split-ring?

22 Emily takes the electric circuit shown in question 19. She changes the battery and switch for a sensitive ammeter. The zero on the ammeter is in the middle of the scale. What happens if she:
(a) pulls the wire out of the magnetic field;
(b) pushes the wire back into the field;
(c) moves the wire faster?

23 Copy and complete the following:
(a) Voltages will be _____ in a wire if the wire moves through a _____ field, or if the field moves past the wire.
(b) This effect is called _____ induction.
(c) The induced voltages will be _____ if the wire cuts the field faster.
(d) Induced _____ tend to oppose the motion.

30
Cathode ray tubes

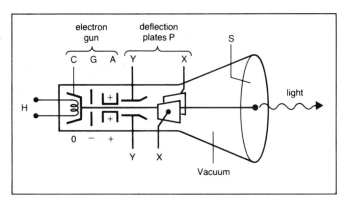

Figure 30.1 *The cathode ray tube (CRT):*

- H = filament **heater**. It needs about 6 V.
- C = the metal-oxide coated **cathode**. It produces lots of electrons by **thermionic emission.**
- A = the metal **anode**. It is held at a few thousand volts positive above the cathode. This high voltage attracts and accelerates the electrons to a high speed.
- G = the **grid**. This controls the number of electrons in the beam. Hence it also controls the brightness of the spot on the screen. When the grid is negative it repels some of the electrons back to the cathode. This reduces the brightness.
- P = **deflection plates**. Positive and negative voltages on the pair of **Y-plates** (Y) cause the beam of electrons to be deflected up or down. Voltages on the **X-plates** (X) cause sideways deflections.

  Deflections by both *X*- and *Y*-plates allow electrons to reach any point on the screen.
- S = the **screen**. The phosphor coating on the inside of the screen converts the kinetic energy of the electrons into light energy.

Electrons

Electrons are found in the outer regions of all atoms. They sometimes behave as separate particles:
- in electric currents in conductors such as metals and in semiconductors;
- in gases conducting electricity;
- in cathode rays from electron guns;
- when emitted as beta particles in radioactive decay.

Electrons have a **negative** electric charge ($-e$). It is equal to -1.6×10^{-19} coulombs.
They have a very small mass. This is written as m_c.

$$m_c = \frac{\text{the mass of a proton or a hydrogen atom}}{1836}$$

$m_c = 9 \times 10^{-31}$ kg which is very small indeed!

What is thermionic emission?

Electrons are normally held captive by their attraction to the positive nuclei of atoms. However:

> **When a metal is heated, some of its electrons gain enough energy to escape from its surface. This is called thermionic emission.**

The following devices use thermionic emission as the source of free electrons:
- the cathode ray tubes (CRTs) in TV sets;
- the CRTs in oscilloscopes and radar monitors;
- X-ray tubes.

The television receiver

The electron gun

An **electron gun** is used to form a narrow beam of fast electrons or a 'cathode ray'. The beam is formed inside an evacuated tube or CRT. The gun has a thermionic source of electrons. This consists of a heated cathode, a grid and one or more anodes.

The television CRT

A television CRT uses coils to deflect the electron beam rather than plates. The beam is deflected at 90° to the magnetic field and at 90° to itself. This is the same as in the 'motor effect'.

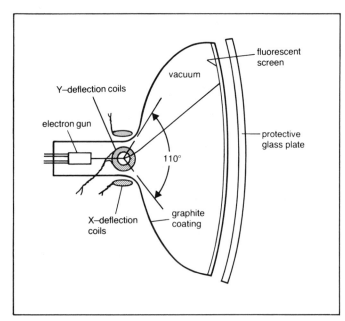

Figure 30.2 *A television CRT:*

The coils are mounted outside the neck of the tube. So they can be fitted and adjusted while the set is being assembled and tested. The strong field of the coils deflects the beam by an angle as wide as 110°. This gives a wide screen for a short tube.

The television picture

- The electron beam is made to scan the whole of the screen. It does this by tracing out 625 lines across it. Alternate lines are traced on each scan of the screen. As a result the picture is renewed completely 25 times a second. But the screen is scanned 50 times a second to reduce the flickering effect.

- As the electron beam scans the screen, varying voltages on the grid in the electron gun control the number of electrons in the beam. By this means, the brightness is varied at different points on the screen to build up a picture.

- Colour pictures are produced by using three different types of phosphor dots on the screen. Three electron guns fire electrons through a mask so that the dots hit by one gun emit red light, by another green light and by the third blue light.

How do we use a cathode ray oscilloscope?

- **Focus and brilliance controls.** These are adjusted to produce a sharp clear spot or trace on the screen.
- **Shift controls.** The X-shift control moves the spot or trace sideways. The Y-shift moves it up and down.
- **Input selector switch: a.c./d.c./earth.** In the d.c. position all voltages, both d.c. and a.c., are displayed on the screen. In the a.c. position only a.c. voltages are displayed. In the 'earth' or 'ground' position the zero of the vertical voltage scale can be set, for example, by moving the spot to the centre or bottom line of the screen (see figure 30.3).
- **Y-gain in volts/division.** This is an amplifier gain control. It is used to make the vertical height of the trace on the screen big enough to measure or study.
- **X-gain and time-base controls.** When waveforms are to be displayed, the CRO draws a voltage–time graph. To get a time scale along the X-axis the spot is made to travel at a constant speed across the screen from left to right. At the right edge it flies back quickly to the left edge. The process is repeated continually.

Measuring voltages on a CRO

A **d.c. voltage** moves the spot up or down to a new fixed position (see figure 30.3 b and c):

$$\frac{\text{d.c.}}{\text{voltage}} = \frac{\text{no. of divs}}{\text{spot has moved}} \times \frac{\text{Y-gain control}}{\text{setting in volts/ div}}$$

An **a.c. voltage** draws a vertical line as it sends the spot up and down rapidly (see figure 30.3 d).

$$\frac{\text{peak-to-peak}}{\text{a.c. voltage}} = \frac{\text{length of vertical}}{\text{line in divs}} \times \frac{\text{Y-gain control}}{\text{setting in volts/div}}$$

- peak-to-peak voltage $= 2 \times \dfrac{\text{peak voltage}}{\text{or amplitude}}$

- The value of an a.c. current or voltage can be measured by a.c. ammeters and voltmeters. They give a kind of 'average' value. This value is called the root mean square (r.m.s.) value.

$$\text{r.m.s. value} = \frac{\text{the peak value}}{\sqrt{2}}$$

How do we display a waveform on a CRO?

Switching the time-base on draws a horizontal line across the screen at a speed selected on the time-base control (see figure 30.3 e).

A waveform can be displayed if there is an a.c. voltage input (see figure 30.3 f).

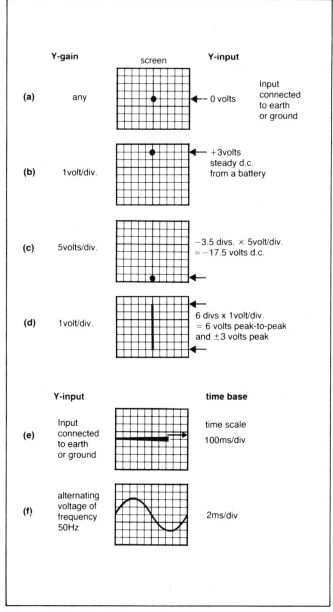

Figure 30.3 *The CRO as a voltmeter.*

Circuit devices and sensors

Fixed resistors

Fixed resistors of maximum power ratings $\frac{1}{4}$, $\frac{1}{2}$, 1 or 2 watts are usually made of carbon or tin oxide in a cylindrical shape. They are colour-coded with bands as shown in figure 31.1.

The maximum power rating must be selected to suit the current to be carried. For example, if a $33\,\Omega$ resistor must carry a current of $0.1\,A$:

$$\text{power} = I^2 R = (0.1)^2 \times 33 = 0.33 \text{ watts}$$

A $33\,\Omega$ resistor rated at a maximum power of $\frac{1}{2}$ watt must be used.

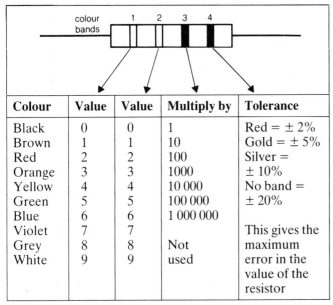

Colour	Value	Value	Multiply by	Tolerance
Black	0	0	1	Red = $\pm 2\%$
Brown	1	1	10	Gold = $\pm 5\%$
Red	2	2	100	Silver =
Orange	3	3	1000	$\pm 10\%$
Yellow	4	4	10 000	No band =
Green	5	5	100 000	$\pm 20\%$
Blue	6	6	1 000 000	
Violet	7	7		This gives the
Grey	8	8	Not	maximum
White	9	9	used	error in the value of the resistor

Figure 31.1 *Colour bands.*

Examples are:
Brown, black, red = 1000 = 1000 ohm = $1\,k\Omega \pm 20\%$
Orange, orange, orange, red = 33 000 = 33 000 ohm
= $33\,k\Omega \pm 2\%$
Green, blue, green, silver = 5 600 000 = 5 600 000 ohm
= $5.6\,M\Omega \pm 10\%$
Brown, green, black, gold = 15 and no noughts
= 15 ohm $\pm 5\%$

Values	Tolerances
R = ohms	G = $\pm 2\%$
K = thousand ohms	J = $\pm 5\%$
M = million ohms	K = $\pm 10\%$
	M = $\pm 20\%$

Examples are:
O R 5 G = $0.5\,\Omega \pm 2\%$
1 R O G = $1.0\,\Omega \pm 2\%$
33 R J = $33\,\Omega \pm 5\%$
K 56 M = $0.56\,k\Omega$ or $560\,\Omega \pm 20\%$
1 K O K = $1.0\,k\Omega$ or $1000\,\Omega \pm 10\%$
M 10 J = $0.10\,M\Omega$ or $100\,000\,\Omega \pm 5\%$

How do we use a variable resistor or rheostat?

Variable resistors are used in circuits to control or vary the current.

● A rheostat can be used to control the brightness of a lamp.
● When the sliding contact B (in figure 31.2a) is moved to:
  end A, zero R gives maximum I;
  end C, maximum R gives minimum I.
● A variable resistor or rheostat has only two connections to a circuit.

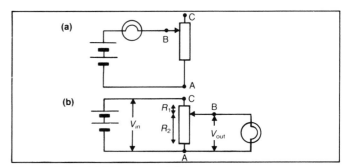

Figure 31.2 *(a) Rheostat and (b) potentiometer circuits.*

How do we use a potentiometer or voltage divider?

A potentiometer (or 'pot') takes a fixed input voltage and divides it into two parts which can be smoothly varied.

Voltage dividers (figure 31.2b) are very common in electronic circuits. They are used as volume, brightness and tone controls.
● The pot or voltage divider uses all three terminals.
● The fixed input voltage is connected across the whole resistor from A to C. This voltage is divided into two parts across R_1 and R_2.
● The output voltages across R_2 can be varied smoothly by moving the sliding contact at B.
● The output voltage can have any value from zero to the full input voltage. It is given by:

$$V_{\text{out}} = V_{\text{in}} \left[\frac{R_2}{R_1 + R_2} \right]$$

Example: A 12 volt battery is connected across a $1\,\text{k}\Omega$ potentiometer. The sliding contact divides the resistance into $800\,\Omega$ and $200\,\Omega$. What is the output voltage across the $200\,\Omega$?

$$V_{\text{out}} = V_{\text{in}}\left[\frac{R_2}{R_1 + R_2}\right] = 12\left[\frac{200}{800 + 200}\right] = 2.4 \text{ volts}$$

Figure 31.3 *The light-dependent resistor, LDR:*

Cadmium sulphide is one of several semiconductor materials whose resistance varies with the amount of light falling on it. Light energy absorbed by the material releases extra charge-carriers. These make it a better conductor. The graph in figure 31.4 shows how the resistance of an LDR (ORP12) varies with illumination (in lux).

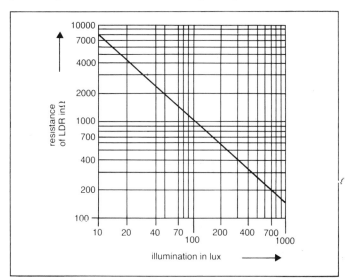

Figure 31.4 *The characteristic of an LDR:*

In bright light of 700 lux, $R = 200\,\Omega$. At 200 lux, $R = 600\,\Omega$. In poor light of 10 lux, $R = 8\,\text{k}\Omega$ or $8 \times 10^3\,\Omega$. In darkness R is about $10\,\text{M}\Omega$.

A simple light meter

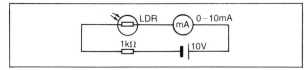

Figure 31.5 *A simple light meter:*

Using the values from the graph above, in bright light the resistance in the circuit is:

$$R \text{ of LDR} + 1\,\text{k}\Omega = 200\,\Omega + 1\,\text{k}\Omega = 1200\,\Omega$$

The milliammeter will read:

$$I = \frac{V}{R} = \frac{10}{1200} = 0.0083 \text{ A or } 8.3\,\text{mA}$$

Figure 31.6 *The thermistor:*

The resistance of most thermistors falls as their temperature rises. Thermistors can be used as temperature sensors to record temperatures. Or they can switch electronic circuits. Examples are thermostats and fire alarms.

The semiconductor diode

The most basic property of a diode is that it conducts in one direction with a low resistance. This direction is called **forward biased**. In the other direction it has a very high resistance. This direction is called **reverse biased**. The arrow in the symbol shows forward bias current flow.

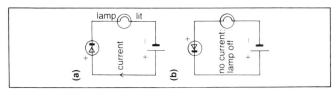

Figure 31.7 *(a) Forward- and (b) reverse-biased diodes.*

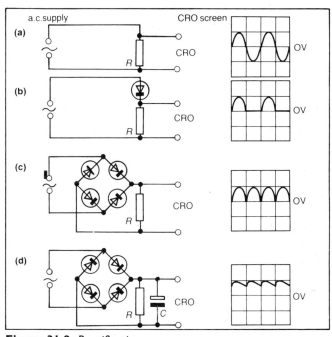

Figure 31.8 *Rectification:*

Rectification is the process of converting alternating current (a.c.) into direct current (d.c.).

(a) With no diode, the voltage output across the load resistor R equals the a.c. input voltage.

(b) With one diode, the output voltage is **half-wave rectified**. This is because the diode allows the current to flow through in one direction only.

(c) With four diodes, the output voltage is **full-wave rectified**.

(d) With four diodes and a capacitor C, the output voltage is partly smoothed to give an almost steady d.c. voltage.

32

Transistor circuits

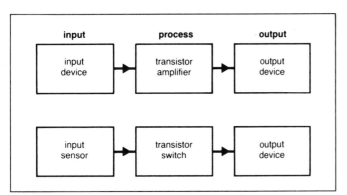

Figure 32.1 *The transistor in a system.*

Input devices and sensors	Output devices
Light sensor (LDR)	Filament lamp
Temperature sensor (thermistor)	Light-emitting
Pressure sensor	diode (LED)
Potentiometer	Warning device
Sound switch or microphone	or buzzer
Rain sensor	Solenoid or relay
	Loudspeaker
	Electric motor

How does a transistor amplify current?

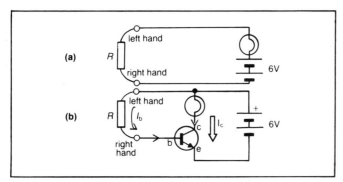

Figure 32.2 *The transistor is a current amplifier:*

b = base; c = collector; e = emitter.

(a) Your body's resistance (R) limits the current to less than 1 mA. So the bulb will not light.

(b) The transistor amplifies the small current through your body. The amplified collector current is large enough to light the lamp.

Why do wet hands make the lamp brighter?

Two currents flow in a transistor:
- a small base current I_b from base to emitter;
- a larger collector current I_c from collector to emitter.

The collector current is an amplified copy of the base current and is controlled by it.
- The amplification is given by:

$$\text{current gain} = \frac{\text{collector current}}{\text{base current}} = \frac{I_c}{I_b}$$

Example: A transistor has a current gain of 80. If the base current is 2.5 mA, what collector current will flow?

$$\begin{array}{c}\text{collector}\\\text{current}\end{array} \quad I_c = \begin{array}{c}\text{base}\\\text{current}\end{array} I_b \times 80 = 200\,\text{mA}$$

How does a transistor work as a switch?

A transistor is switched on and off by the voltage between its base and emitter (V_{be}). When the voltage on the base is:
- *less* than 0.6 volts, the transistor is switched *off* and *no* collector current flows;
- *greater* than 0.6 volts the transistor is switched *on* and a collector current (I_c) flows.

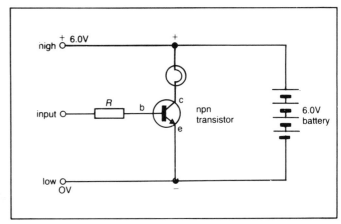

Figure 32.3 *The transistor is a switch:*

The base resistor R protects the transistor. It does this by limiting the base current. The input is made 'high' by connecting it to the +6 volt power line. It is made 'low' by connecting to zero volts.

Input	V_{be}	Transistor	I_c	Output lamp
Low	< 0.6	Off	Zero	Off
High	> 0.6	On	Flows	On

Advantages include:
- A transistor switch can be operated by many different inputs. Examples are: sound, light, temperature, pressure or rain.
- It can switch very quickly.
- It has no moving parts to wear out.

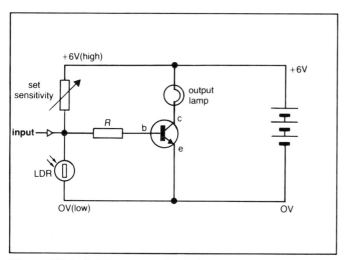

Figure 32.4 *A light-operated switch:*

When a bright light falls on the LDR its resistance will fall to a few hundred ohms. The voltage across it will be small. This makes the input to the transistor low and turns it off. In dim light the resistance of the LDR increases to several thousand ohms. The larger voltage across the LDR turns the transistor on and the output lamp lights.

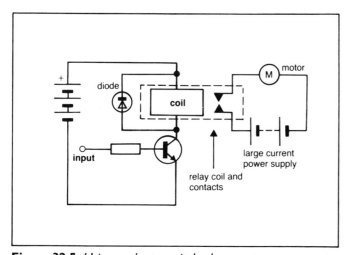

Figure 32.5 *Using a relay to switch a large current:*

Currents of several amperes would overheat and destroy most transistors. So to switch a large current a relay is used which is controlled by a transistor.

● When the transistor is turned on its collector current flows through the relay coil.
● The relay coil is an electromagnet. It can switch a much larger current in a separate circuit. It does this by opening or closing contacts.
● A diode is needed across the relay coil. The diode must be connected in the reverse bias direction. This diode protects the transistor from the large voltage induced across the relay coil when the current through it is switched off. The diode protects the transistor by shorting out the voltage.

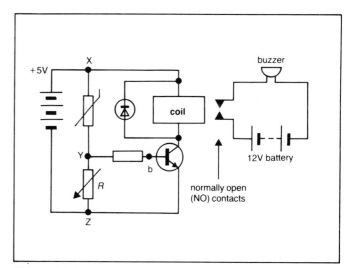

Figure 32.6 *A temperature-operated switch:*

NO = normally open contacts. The thermistor and R form a voltage divider. When the thermistor gets hot its resistance falls. So the voltage across it (XY) also falls. This means the voltage across R (YZ) rises. This is because voltage XY + voltage YZ = voltage XZ. This is equal to the constant voltage of the 5 V power supply. As the voltage YZ increases the voltage on the base b also increases. The transistor switches on when the base voltage exceeds 0.6 volts. The collector current now flows and magnetises the relay. This connects the buzzer to the 12 V power supply. This circuit makes a fire alarm.

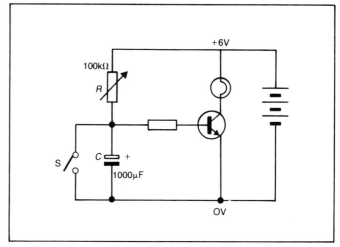

Figure 32.7 *Time-delayed switching:*

Closing switch S discharges (empties) the capacitor C. Opening switch S starts the time-delayed switching. The battery charges the capacitor through the variable resistor R. When the voltage across the capacitor C reaches just over 0.6 volts it is high enough to switch on the transistor. Both C and R determine the time taken to reach 0.6 volts. A larger capacitor takes longer to fill. A smaller current flows in a larger resistor.

33

The operational amplifier

What are analogue signals?

An analogue signal in an electronic system is one which:
● is in the form of a voltage or current;
● can vary smoothly and continuously;
● is an electrical model (analogue) of a physical quantity such as pressure, temperature, position or speed.

We get analogue signals from: a tape recorder, thermometer, light meter, force meter, microphone or barometer.

Both the amplitude (loudness) and the frequency (pitch) of a sound can vary smoothly.

Operational amplifiers process analogue signals.

What is an operational amplifier?

An operational amplifier or 'op-amp' is a complex circuit containing many transistors and resistors.

The basic circuit has the following properties:

1. It has a very high voltage gain – up to 100000.
2. It has very high input resistance – millions of ohms.
3. It has a much lower output resistance – a few hundred ohms.
4. It can amplify both d.c. and a.c. signals.
5. The input and output voltages can be either positive or negative.
6. The output voltage cannot exceed the power supply voltages (typically + 15 and − 15 volts).

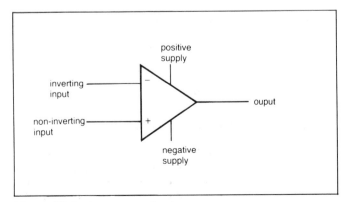

Figure 33.1 *An op-amp circuit:*
The symbol shows that it has two inputs labelled − and + .

The inverting input (−) turns a positive input voltage into a negative output voltage. It also *inverts* the waveform of an alternating signal.

The non-inverting input (+) does not alter the sign of the input voltage nor invert an alternating signal.

The op-amp needs both a positive and negative power supply. It also needs an earth or zero volts line. The power supply connections are often omitted from simple circuit diagrams.

What is feedback?

When part of the output of a system is fed back into its input we say there is feedback.

Positive feedback occurs when adding the output to the input makes the output even bigger. Positive feedback can lead to an oscillation. An example is when the output of a public address system is fed back into the microphones.

Negative feedback occurs when adding part of the output to the input makes the output smaller. Negative feedback is usually used with amplifiers because:

● It reduces distortion. This gives a clearer signal.
● It makes the voltage gain of the amplifier accurately predictable and independent of the particular transistor or integrated circuit used.
● It makes the voltage gain constant over a wide range of frequencies (a wide bandwidth).

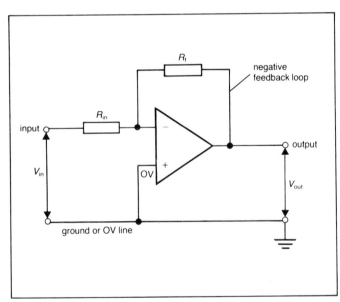

Figure 33.2 *An inverting amplifier using negative feedback:*

The inverting input is used. So the output voltage will be of *opposite* sign to the input voltage. Some of the output is fed back to the input through the **feedback resistor R_f.** This feedback signal will subtract from the input voltage because it is of opposite sign. This *reduces* the final output voltage giving negative feedback.

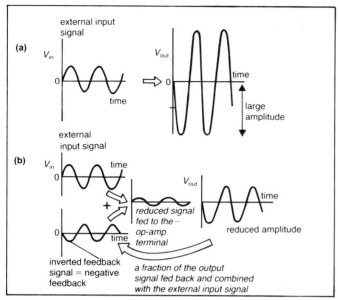

Figure 33.3 *The effect of negative feedback on an alternating signal such as a sound wave:*

(a) Without negative feedback V_{out} has a large amplitude and is inverted.

(b) With negative feedback V_{out} is still inverted but the amplitude is reduced.

How do we calculate the voltage gain?

The complete amplifier including the input resistor R_{in} and the feedback resistor R_f has a voltage gain A given by:

$$\text{amplifier voltage gain } A = - \frac{V_{out}}{V_{in}}$$

- The minus sign means that the output is inverted or of the opposite sign to the input.
- V_{in} and V_{out} can be measured by connecting an oscilloscope as shown in figure 33.3.
- Because negative feedback is used, the voltage gain is fully controlled by the values of the resistors R_{in} and R_f.
- The voltage gain A can be calculated using:

$$\text{voltage gain } A = - \frac{R_f}{R_{in}}$$

Example: An inverting op-amp has a feedback resistor of $200\,k\Omega$ and an input resistor of $10\,k\Omega$. It receives a signal of $+0.4$ volts. Calculate the output voltage.

$$\text{voltage gain } A = - \frac{R_f}{R_{in}} = - \frac{200\,k\Omega}{10\,k\Omega} = -20$$

The output voltage = input voltage $\times -20$:

$$\therefore \text{ output voltage} = 0.4 \times -20 = -8.0 \text{ volts}$$

What is saturation?

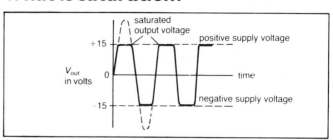

Figure 33.4 *Output voltage saturation:*

The output voltage saturates when the output voltage tries to exceed the power supply voltages ($+15$ and -15 volts). This happens when:

- the input voltage is too large; or
- the voltage gain is too large.

Saturation is the *flattening* or *clipping* effect which stops the output voltage reaching its full amplified value.

The dotted curve shows what the output would be if it were not limited by the supply voltages of the op-amp.

What is the frequency response of an amplifier?

Using the inverting amplifier with negative feedback the voltage gain can be measured over a wide range of frequencies. The input signal is obtained from a signal generator.

Frequencies of 1 hertz to 1 megahertz rising in steps of times ten cover the range of a typical op-amp. Figure 33.5 shows a typical **frequency response** when the voltage gain has been set at $\times 100$. From 1 Hz to about 10 kHz the voltage gain is constant at $\times 100$. This frequency range is called the **bandwidth**.

At higher frequencies the gain gets smaller.

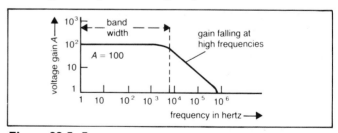

Figure 33.5 *Frequency response.*

What is biological feedback?

Close your eyes. Bring your two index fingers together from some distance apart. Point them so that they should touch at the tips. Try this several times. How often do you miss? Why do you think you miss? You need negative feedback through your eyes to your brain to correct the errors in your arm movements.

Review questions: Chapters 30, 31, 32 and 33

C30 1 (a) What type of beam does a cathode-ray tube produce?
   (b) Is the beam produced at the anode or cathode?
   (c) What is produced when the beam hits the screen?
   (d) Which plates move the beam up and down?

2 (a) Are electrons positively or negatively charged?
   (b) Which is the charge of an electron:
       (i) 1.6×10^{19} C; (ii) -1.6×10^{19} C;
       (iii) -1.6×10^{-19} C; (iv) 1.6×10^{-19} C?
   (c) Is an electron's mass smaller than, equal to, or greater than a proton's mass?

3 (a) What is meant by *thermionic emission*?
   (b) Which of the following use thermionic emission to produce free electrons?
       (i) X-ray tube; (ii) Geiger–Müller tube;
       (iii) cathode-ray tube; (iv) battery?

4 (a) What is used to deflect the beam of electrons in a television tube?
   (b) How many lines are traced on the screen?
   (c) Why is the screen scanned 50 times per second?
   (d) Which three colours of light are emitted by the phosphor dots in a colour TV set?

5 The figure shows traces on an oscilloscope screen. (A) shows the spot set for an input of 0 volts. The volts/division is set at 2 V/div.

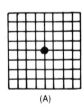
(A)

(B)

(C)

(D)

   (a) What is the input voltage for (B)?
   (b) What would be the effect of changing around the input connections?

   (c) Describe the input signal in (C).
   (d) Which oscilloscope control has been switched on to give the trace in (D)?
   (e) What is the peak-to-peak voltage in (D)?

C31 6 A boy wishes to use a $100\,\Omega$ resistor to carry currents of up to 0.1 A. What power rating should the resistor have?

7 What are the values of resistors with the following coloured bands:
   (a) Red red orange gold;
   (b) Brown black brown silver?

8 What are the values of resistors with the following letter codes:
   (a) 4 K 7 J; (b) M 33 M; (c) 56 R K; (d) 10 K J?

9 What are the: (i) colours; (ii) letter codes of the following resistors:
   (a) $6.8\,k\Omega \pm 10\%$; (b) $470\,\Omega \pm 5\%$; (c) $820\,k\Omega$?

10 Copy and complete the following:
   (a) A *rheostat* is used to vary the _____ flowing in a circuit.
   (b) A *potentiometer* is used to give a variable _____ supply.
   (c) A *potentiometer* is sometimes called a _____ divider.

11 A 12 V battery is connected across a $100\,\Omega$ potentiometer as shown in the figure. What is the output voltage across R_2 if:
   (a) $R_1 = R_2$;
   (b) $R_1 = 2R_2$;
   (c) $2R_1 = R_2$;
   (d) $R_1 = 25\,\Omega$ and $R_2 = 75\,\Omega$;
   (e) $R_1 = 30\,\Omega$ and $R_2 = 70\,\Omega$?

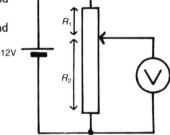

12 (a) What happens to the resistance of an *LDR* as the light falling on it increases?
   (b) As a sensor, what can a *thermistor* be used to detect?
   (c) In which direction do *diodes* have a very low resistance?
   (d) What is the name of the process for converting a.c. into d.c.?
   (e) Which device is used to convert a.c. into d.c.?
   (f) What are the symbols of:
       (i) LDR; (ii) thermistor; (iii) diode?

C32 **13** Which of the following are output devices:
(a) LDR; (b) microphone; (c) relay; (d) motor;
(e) transistor; (f) buzzer; (g) filament lamp?

14 (a) How many legs has a transistor?
(b) In a transistor, what do the letters b, c and e stand for?
(c) What are the names of the currents which flow into a transistor?

15 A transistor has a current gain of 10. What is the value of:
(a) I_b if $I_c = 60\,mA$;
(b) I_b if $I_c = 0.02\,A$;
(c) I_c if $I_b = 10\,\mu A$;
(d) I_c if $I_b = 5\,mA$?

16 Copy and complete the following:
(a) In a transistor switch circuit, the transistor is turned _____ when the input voltage is low.
(b) When a transistor is _____ the collector current I_c flows.

17 The figure shows part of a transistor switch circuit.
(a) Use the letters on the diagram to say where the following should be connected:

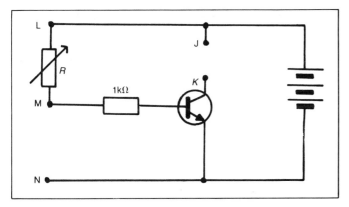

(i) an output lamp;
(ii) an input sensor.
(b) What input sensor is used to make:
(i) a light-sensitive circuit;
(ii) a temperature-sensitive circuit?
(c) With the input sensor(s) in this position, what input conditions will cause the transistor to switch on?
(d) How can the circuit be altered to switch the transistor on in the opposite input conditions?
(e) Which component is used to alter the sensitivity of the circuit?

18 (a) What device is needed in a transistor switch circuit to operate a heater?
(b) Why is this device needed?
(c) Why is a diode needed across this device?

C33 **19** Copy and complete the following:
(a) An analogue signal in an electrical circuit is in the form of a _____ or _____.
(b) Analogue signals can vary smoothly and _____.

20 (a) What is a typical value for the voltage gain of an op-amp with no feedback?
(b) Is the *input* resistance of an op-amp very high or very small?
(c) Can an op-amp amplify both a.c. and d.c. signals?
(d) What are the names of the inputs labelled + and −?

21 (a) Which type of feedback makes the output get bigger and bigger?
(b) What effect does *negative* feedback have on the distortion of a signal?
(c) If an amplifier is described as having a wide bandwidth, what does this mean about the voltage gain of the amplifier?

22 The figure shows an op-amp circuit. Assume the power supply is ± 15 V.

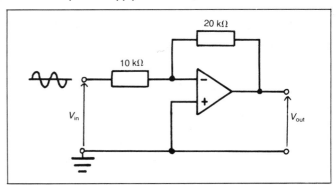

(a) What is the name of the input terminal of the op-amp in this circuit?
(b) What type of amplifier circuit is shown?
(c) What size is the:
(i) input resistor;
(ii) feedback resistor?
(d) What is the voltage gain of this circuit?
(e) If the input signal has a peak value of ±5 V, what is the peak value of the output?
(f) Describe how the output signal will look different from the input signal.
(g) Repeat part (e) for an input of ± 10 V.

23 (a) What does *saturation* mean in amplifiers?
(b) What is a graph of voltage against frequency called?
(c) What is constant when the graph is flat?
(d) What is the frequency range called where the graph is flat?
(e) What happens to the voltage gain at high frequencies?

34
Digital electronics

What is digital information?

Any information which has a limited number of values and can change only suddenly from one value to another is digital information.

● We get digital signals or information from: central heating controls (off or on), a combination lock (locked or open), a computer memory and warning lights on a car dashboard.

What is a binary digital system?

The simplest digital systems use only *two* states. The two states can be stored or processed in a variety of forms. For example, the table shows several ways of giving the answer to the question: 'Is the door closed?'

Answer **yes**	Closed switch	Magnetised memory	High voltage	Logic 1
Answer **no**	Open switch	Memory not magnetised	Low voltage	Logic 0

● In digital electronics logic 1 can be shown by a 5 volt signal. Logic 0 can be shown by zero volts.
● The logic 0 and 1 states are used as 0 and 1 in binary arithmetic.

Logic gates are the decision-making units of a digital system.

Output devices

Logic gates in the form of integrated circuits (ICs) can supply only a limited output current. This can drive a light-emitting diode (LED). Or it might power a suitable relay.

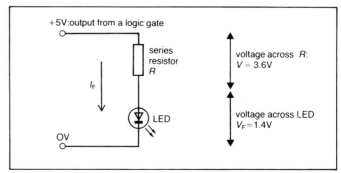

Figure 34.2 *LED output indicator.*

Logic gate	Symbol	Is equivalent to	Truth table			The output is high, logic 1 when:
NOT	A ▷○ Y	INVERTER	input		output	Input A is **NOT** high (output is the input inverted)
			A		Y	
			0		1	
			1		0	
OR	A, B ⊃ Y	(inclusive) OR	A	B	Y	Input A **OR** input B is high (or both are high)
			0	0	0	
			0	1	1	
			1	0	1	
			1	1	1	
NOR	A, B ⊃○ Y	OR-NOT	A	B	Y	Neither input A **NOR** input B is high
			0	0	1	
			0	1	0	
			1	0	0	
			1	1	0	
AND	A, B ⊐ Y		A	B	Y	Input A **AND** input B are high
			0	0	0	
			0	1	0	
			1	0	0	
			1	1	1	
NAND	A, B ⊐○ Y	AND-NOT	A	B	Y	Input A **AND** input B are **NOT** both high
			0	0	1	
			0	1	1	
			1	0	1	
			1	1	0	

Figure 34.1 *Logic gates.*

- A LED is a diode. So it must be forward biased.
- A LED uses only a small current and low power.
- A LED needs a resistor in series with it to limit the current through it.

Example: A LED is to be used as an indicator for the output of a logic gate. It has the following ratings.
Power supply = 5.0 volts.
V_F (suitable forward voltage): 1.4 volts
I_F (typical forward current): 20 mA or 0.02 A
Calculate the resistance of the series resistor R needed in series with this LED.

$$\text{voltage across series resistor} = 5.0 - 1.4 = 3.6\,\text{V}$$

$$\therefore \text{series resistance } R = \frac{V}{I_F} = \frac{3.6\,\text{V}}{0.02\,\text{A}} = 180\,\Omega$$

Relay output device

- A relay coil can work on the small current available from the output of a logic circuit.
- The relay contacts can switch a much larger current in another separate circuit.

Combinations of logic gates

Logic gates are often combined together:
- to make more complicated decisions than the single gates (AND, OR, NAND or NOR) and to build adding, counting and memory systems;
- to allow a system to be built from entirely one kind of gate (e.g. NAND gates). This is because many NAND gates can be made in a single chip.

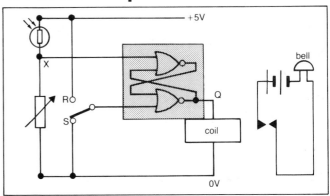

Figure 34.3 *Combinations of logic gates:*

(a) NAND; (b) NAND; (c) NOT; (d) AND; (e) inclusive OR. The inclusive OR includes the case when A and B are high.

How do we design a simple system?

Look for the words AND, OR, NOR and NOT. Connect gates together to fit these words.

Example: Draw a truth table for a system which will sound an alarm when it is raining *or* windy *and* the alarm is switched on. Draw a logic gate combination for this system.

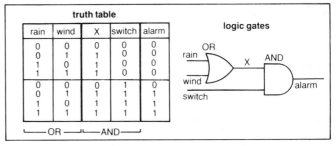

Figure 34.4.

What is a simple latch?

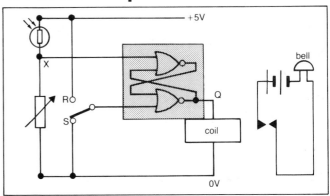

Figure 34.5 *A simple latch or memory unit:*

The pair of linked NOR gates in the circuit is called a **bistable latch**.

> **A bistable latch 'latches' its output Q at either 0 or 1 and stays there until its other input is changed or reset.**

First the switch is down (logic 0). When a light shines on the LDR its resistance falls. Point X goes to a high voltage (logic 1). The output Q changes to logic 1. It turns on the alarm. The latch circuit stays in this state with the alarm on until the switch is reset. It does this even if the light goes off. The circuit 'remembers' the shining light.

Communication systems

What happens in a communication system?

The information to be transmitted is called the signal.

So that the signal is not lost, it must remain large compared with any unwanted **noise** such as hiss and crackle. The ratio between the amplitude of the signal and the noise is called the **signal-to-noise ratio.**

The signal loses power as it travels. This is called attenuation.

Amplifiers are needed to boost weak signals and improve the signal-to-noise ratio.

The encoding process is how the information is prepared for its journey. When it is added to a carrier wave the process is called modulation.

The information **receiver** may be a person, a computer, a printer or other controlled machine.

What types of modulation are used?

Amplitude modulation (AM)

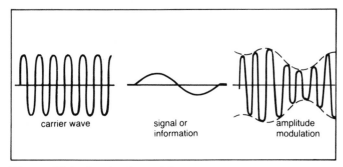

Figure 35.1 *Amplitude modulation:*

The radio carrier wave has a *constant frequency* somewhere in the range 0.1 to 10 MHz. The signal, of much lower frequency, (hearing range: 20 Hz to 20 kHz) is added to the carrier wave. This has the effect of varying the amplitude of the carrier wave in a way which matches the signal.

Frequency modulation (FM)

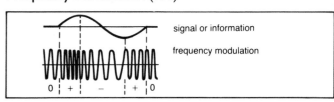

Figure 35.2 *Frequency modulation:*

In this case the radio carrier wave has a *constant amplitude* but its frequency varies. The frequency increases (+) and decreases (−) as the amplitude of the signal varies.

Pulse code modulation (PCM)

In PCM the amplitude of the signal is *sampled* at regular intervals typically 8000 times per second. The sample is then given a digital value in the form of an eight-bit binary number. (Values from 0 to 255 are possible: binary 11111111 = 255.)

The binary numbers are encoded as voltage pulses. These have a positive value to represent 1. Zero volts represents 0.

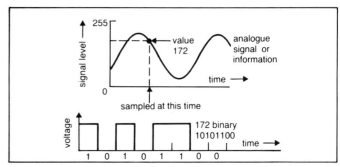

Figure 35.3 *A string of binary-coded pulses gives the value of the signal at a particular moment:*

The carrier wave is modulated with a string of these voltage pulses. These are a binary-coded form of the signal (PCM).

The main advantage of PCM is that the decoding device at the receiver has only to detect whether there is a pulse or not. The shape of the pulse does not matter. So the added noise has no effect as long as the pulse can be 'seen'.

What types of transmission links are used?

The following kinds of links are used:

① Coaxial cables, across land and sea.

② Optical fibres, also across land and sea.

③ Microwave links directly between towers.

④ Satellite microwave links.

⑤ Very high frequency (VHF) and ultra high frequency (UHF) radio and TV broadcasts.

⑥ Sky waves – these are radiowaves reflected from the ionised layers in the upper atmosphere, called the ionosphere.

Radio communication systems

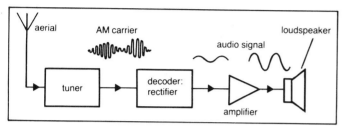

Figure 35.4 *AM radio receiver:*

● A particular station is found by matching a tuner circuit to the frequency of the station. The tuner then rejects all other frequencies.
● A rectifier or detector separates the audio frequency (AF) signal from the radio frequency (RF) carrier wave.
● The amplifier boosts the signal strength.

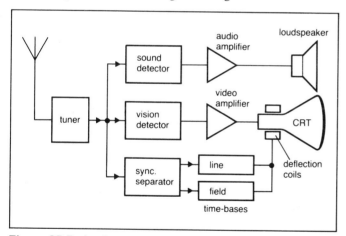

Figure 35.5 *A television receiver:*

● Television signals are transmitted as frequency-modulated (FM) RF waves.
● The RF carrier wave is picked up by an aerial. A particular 'channel' is selected by tuning the tuner to the frequency of the carrier wave for that channel.
● The signal contains three main components. They are separated by electronic circuits.
● The picture is synchronised with the scanning of the electron beam in the CRT by two sets of pulses. These start the **line** and **field time-bases** at the right moments. The line time-base must make the electron beam start to scan across the screen at the right time. The field time base starts the picture at the top of the screen and not somewhere else.

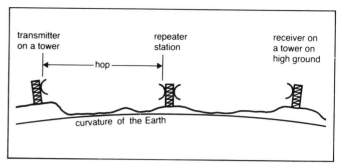

Figure 35.6 *Terrestrial microwave links:*

Concave **dish aerials** are placed on towers or high buildings. They are used to transmit and receive microwaves of super-high frequencies of about $10\,GHz$ (= $10^{10}\,Hz$). A link is made up of several hops because:

● the curvature of the Earth limits the distance over which there is a direct 'line of sight';
● the microwaves get weaker with the square of the distance travelled. So they need amplifying at **repeater stations.**

In the UK much telephone and television communication uses terrestrial microwave links. These links usually use PCM.

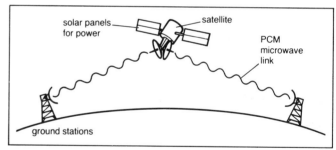

Figure 35.7 *Satellite microwave links:*

Communication satellites are most often placed in a high orbit over the equator.

This orbit is called a geosynchronous or geostationary orbit. In it the satellite orbits the Earth once every 24 hours. So it appears to hover above the same point on the Earth.

Satellites have a large capacity. They can carry thousands of telephone lines and tens of television links at the same time. Once in orbit, satellites are very reliable. These microwave links use frequencies in the 3 to 30 GHz range. They carry digital signals with PCM.

Telephone systems

Telephone systems now carry computer data, text, picture and speech information. To connect the computers to the telephone lines, a modulator–demodulator or 'modem' is needed. It converts between analogue and digital signals at both ends of the line.

36 Materials

How is matter built up?

	Diameter	How many in 1 mm
Atom	10^{-10} m (typically 0.1 to 0.5 nm)	2 000 000 to 10 000 000
Molecule (linked atoms)	10^{-9} m = 1.0 nm	1 000 000

- **All substances are made up of atoms.**
- **In an element all the atoms are identical.**
- **A compound has two or more kinds of atoms.**
- **An atom is the smallest sample of an element.**
- **Atoms of different elements link to form molecules of compound substances.**

What are the mechanical properties of a solid?

- **Strength.** A strong material needs a big force to break it. Strength often depends on whether the material is stretched, compressed or twisted.
- **Stiffness.** A stiff material does not stretch, bend or 'give' very much. It is rigid.
- **Elasticity.** When the force is removed an elastic material goes back to its original size and shape. (It may have been pulled, bent or twisted.)
- **Plastic** materials stay permanently deformed or stretched. This is so even when the force is removed.
- **Ductile** materials can be pulled out into wires or rolled into thin sheets. Metals are ductile.
- **Brittleness.** A brittle material breaks suddenly and is fragile. Bricks and glass are brittle.

The three states of matter

How do materials stretch?

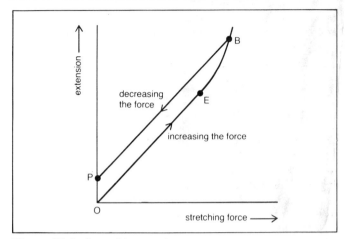

Figure 36.1 *Stretching a spring:*

The extension is the *increase* in length of the spring when it is pulled by the stretching force.

- Extension values are found by subtracting the reading for the unstretched spring from all the scale readings.
- The graph is a straight line from O to E. This shows that the extension increases in proportion with the stretching force up to the **elastic limit** E. This is known as **Hooke's law**.
- From E to B the spring gets a *permanent* stretch equal to OP. This is not elastic.
- A steel wire behaves the same way when stretched.

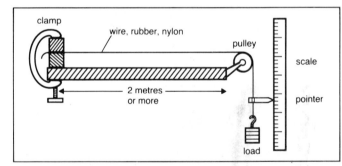

Figure 36.2 *Testing materials:*

State	Spacing of molecules	Volume and shape	Special properties	Movement of molecules
Solid	Close packed in a regular formation giving high density	Fixed volume and shape	Can be cut, stretched, bent, twisted and polished	Limited to vibrations about a fixed position
Liquid	Close packed but without order, also high density	Constant volume, no fixed shape	Can be poured, forms drops, takes the shape of its container	Can move throughout the liquid but cling together
Gas	Far apart giving low density	Neither volume nor shape fixed, compressible	Expands to fill the space available	Independent random motion

- Use a long length of the material to be stretched. This makes the extension larger.
- Take readings for loading and unloading.
- Extension = scale reading when loaded − scale reading for no force.

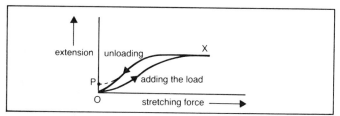

Figure 36.3 *Stretching rubber:*

- Rubber does not obey Hooke's law i.e. the graph is not straight anywhere.
- The rubber tends to 'remember' being stretched. It follows a different curve while being unloaded to that when it was loaded.
- A permanent stretch or 'set' may remain. This is OP on the graph.

Example: Using the graph find:
(a) the extension when the load is 0.7 newtons;
(b) the load needed to produce an extension of 14 mm;
(c) the greatest load which will not stretch the spring beyond its elastic limit.

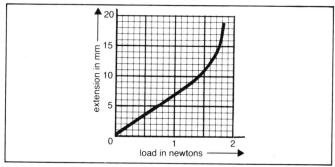

Figure 36.4:

(a) extension = 5.0 mm; (b) load = 1.7 newtons; (c) maximum load for elastic stretching = 1.3 N. This is the point where the graph begins to curve.

What makes a strong beam?

Some materials such as stone and concrete are strong under *compression*. But they are much weaker when *stretched*. Materials such as steel are strongest under tension. Designers of buildings and bridges must take care the correct material to select to suit the forces acting. Increased strength is made possible by:

- Using a beam or cable with a larger cross-sectional area. A thicker cable can bear a larger load before reaching its elastic limit or breaking point.

- Forming the material into I-shaped or box-sectioned girders. This gives more strength for less weight.

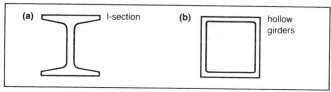

Figure 36.5 *Girder sections:*

(a) An I-section girder is strong. (b) Hollow girders are light and strong.

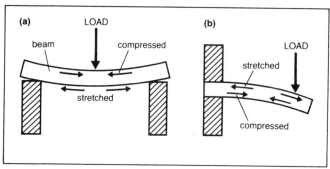

Figure 36.6 *Bending beams:*

(a) Central load. (b) A cantilever. This is a beam supported at one end only.

What makes bridges strong?

Bridges come in many designs. They have to be very strong. Look at the two examples in figure 36.7. They show how materials are chosen for their strength under the kind of compression or tension forces present in the structure.

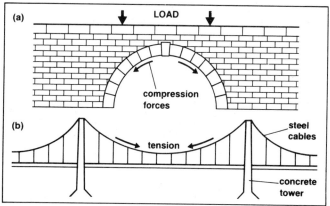

Figure 36.7 *Bridges:*

(a) Stone arch – strong in compression.
(b) Suspension bridge – steel cables are strong in tension. Concrete towers are strong in compression.

Review questions: Chapters 34, 35 and 36

C34 1 Copy and complete the following:
  (a) Digital signals have only a certain _____ of values.
  (b) _____ digital signals have two states. They are called logic 0 and logic 1.
  (c) Digital signals _____ suddenly from one value state to another.

2 Draw the symbol and write truth tables for the logic gates:
  (a) NOT; (b) OR; (c) AND; (d) NOR; (e) NAND.

3 What is the name of the logic gate which:
  (a) is an inverter;
  (b) is equivalent to OR-NOT;
  (c) is equivalent to AND-NOT;
  (d) has only one input;
  (e) gives a high output when all inputs are high?

4 Write truth tables for the combination of gates shown below. State which **one** gate could be used to replace each combination.

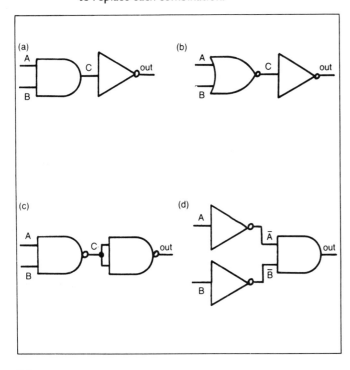

5 Draw a truth table and a logic gate combination for a system which:
  (a) will sound an alarm when it is hot and the alarm is switched on;
  (b) will light a lamp when it is dark or the switch is pressed;
  (c) will turn on a heater when it is cold and either the timer is on or the manual override switch is pressed.

6 (a) What do the letters *LED* stand for?
  (b) What is the electrical symbol for a LED?
  (c) Why must a resistor be put in series with a LED?
  (d) Does a LED use a small or large current?

7 A girl wishes to use a LED as an indicator with a 5.0 V supply. She chooses a LED with the following data: $V_F = 1.7$ V, $I_F = 10$ mA. She knows that she also needs to choose a suitable resistor.
  (a) How will the diode and resistor need to be connected in the circuit?
  (b) What is the voltage across the resistor?
  (c) What is the current through the resistor?
  (d) What value resistor is needed in the circuit?

8 Copy and complete the following:
  (a) A bistable latch is a _____ unit.
  (b) It latches its _____ Q and stays there until its other output is changed or _____.

9 Design a system with a latch to switch on:
  (a) a lamp when it gets dark;
  (b) a heater when the temperature falls.

C35 10 In a communication system:
  (a) What are unwanted crackles called?
  (b) What is attenuation a loss of?
  (c) What is adding a signal to a carrier wave called?

11 For modulated signals:
  (a) What do the letters AM stand for?
  (b) What do the letters FM stand for?
  (c) Is the frequency of the carrier wave much less than or much higher than the frequency of the information signal?
  (d) Which type of modulation has a constant amplitude?
  (e) Sketch graphs to show:
      (i) AM; (ii) FM waves.

12 (a) What do the letters *PCM* stand for?
  (b) In PCM, what is used to give the value of an analogue signal at a particular moment?

13   A man listens to a concert on the radio:
    (a) Are the sound waves AF or RF waves?
    (b) Which part of a receiver selects one radio frequency?
    (c) What does the rectifier do?
    (d) Which waves are heard from the loudspeaker?

14   (a) Are television signals transmitted as AM or FM radio waves?
    (b) What are the three main components of the signal?
    (c) How is the picture synchronised?

15   In terrestial microwave links:
    (a) Are the dish aerials convex or concave?
    (b) Are the microwaves used of high or low radio frequencies?
    (c) Why is a direct 'line of sight' distance limited?
    (d) Why are repeater stations used?

16   What happens to the strength of microwaves from a transmitter if the distance away is:
    (a) doubled; (b) four times bigger?

17   (a) What are satellites which appear to hover above the same point on the Earth called?
    (b) What kind of signals are usually used with the microwaves to satellites?

C36   18   Copy and complete the following:
    (a) In an _____ all the atoms are identical.
    (b) A _____ consists of more than one kind of atom.
    (c) _____ are atoms linked together in a certain way.

19   (a) What are the three states of matter?
    (b) In which state are the molecules:
      (i) very far apart;
      (ii) limited to vibrations about fixed positions;
      (iii) close packed, but without order?
    (c) In which state can a substance be:
      (i) bent; (ii) poured; (iii) polished?
    (d) In which state is the density of a substance low?

20   Choose from the list – copper, plasticine, glass, rubber, cast iron – one which is:
    (a) most elastic; (b) very strong;
    (c) plastic; (d) very brittle; (e) ductile.

21   Copy and complete the following:
    A material which is:
    (a) _____ can be pulled into wires.
    (b) _____ can stay permanently deformed.
    (c) _____ needs a big force to break it.
    (d) _____ is very rigid.

22   A 40 cm long spring stretches to 45 cm when a weight is attached to its end.
    (a) What is the original length of the spring?
    (b) What is the extension of the spring?
    (c) Assume the spring has not reached its elastic limit. What length will it be when the weight is removed?
    (d) Which law will be obeyed?

23   The figure shows an extension–force graph.

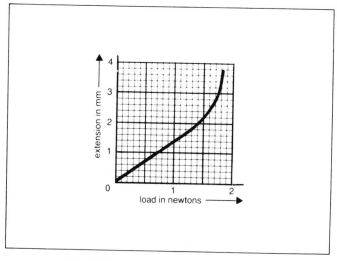

    (a) What is the load needed to give an extension of: (i) 1.0 cm; (ii) 1.4 cm?
    (b) What is the extension for a load of: (i) 1.5 N; (ii) 0.4 N?
    (c) For what maximum extension will the spring be able to return to its original length?

24   The figure shows some beams being bent by loads.

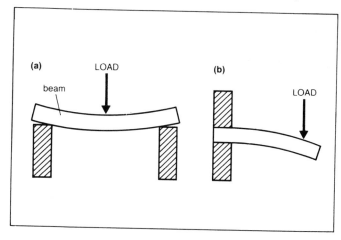

Copy the diagrams and show where the beams are being:
(a) stretched; (b) compressed.

Heat and temperature

Heat is a form of *energy*. When absorbed by an object it makes the object hotter. When lost by an object it leaves it colder. The 'hotness' or 'coldness' of an object we call its temperature.

Temperature is also a measure of the average kinetic energy of the molecules in an object.

What is a temperature scale?

Temperature is measured with a thermometer. The scale on a thermometer is called a **temperature scale**. To make a scale on a thermometer we choose two fixed and easily obtainable temperatures.

The Celsius scale

● **The lower fixed point.** This is known as the **ice point.** It is the temperature of melting pure ice: 0°C.

● **The upper fixed point.** This is known as the **steam point.** It is the temperature of steam just above boiling water: 100°C.

The scale between these two fixed points is divided evenly into 100 degrees. This makes it a centigrade (= 100 graduations) scale.

The Kelvin or absolute scale

The Kelvin temperature scale has its zero at absolute zero. 0 Kelvin = −273°C.

> **Absolute zero is the lowest possible temperature which can exist. It is that temperature at which no more heat can be removed from a substance. Its molecules have no movement.**

The Kelvin scale uses the same fixed points as the Celsius scale. There are also 100 degrees between them.

Conversion between scales

T (kelvin) = t (celsius) + 273

Examples: Wax melts at 57°C.

∴ the melting point of wax = 57 + 273 = 330 K

An object appears red hot at about 1100 K. The temperature of red heat on the Celsius scale is given by:
$t = T$ (kelvin) − 273

∴ t = 1100 − 273 = 827°C

Type of thermometer	Range or use
Mercury-in-glass	Max. range: −39°C to +630°C. Common: −10°C to +100°C in 1° divs. Easy to use and portable
Clinical	35°C to 42°C in 0.1° or 0.2° divs. Measures body temperature
Alcohol-in-glass	Max. range: −117°C to +79°C. Lower range than mercury. Used in very cold places like the Arctic
Thermocouple	Has two metals joined in a circuit. It generates a small voltage when the two junctions are at different temperatures. Can respond quickly and gives a continuous reading
Thermistor	A resistor whose resistance varies with temperature. In an electric circuit can give continuous readings

Expansion

When an object is heated its molecules vibrate more violently. This is because they have more kinetic energy. They also need more space around them. This causes the material to expand.

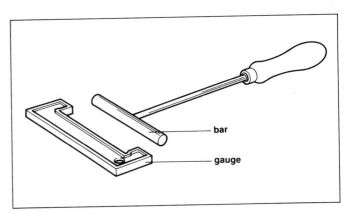

Figure 37.1 *Bar and gauge:*

The bar fits the gauge when both are cold. When the bar has been heated it no longer fits. The expansion is too small to see without the gauge.

● The expansion of solids is very small. But the force of expansion is very strong.
● The expansion of liquids and gases is much greater. This is because the molecules are free to move.

The bimetallic strip

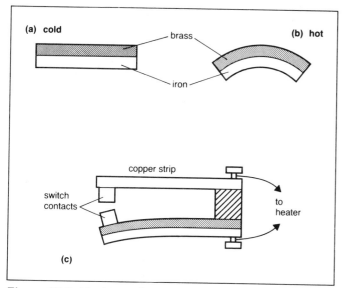

Figure 37.2 *The bimetallic strip:*

It is made of two strips of different metals joined together. Examples are brass and iron.

(a) When cold the double strip is straight.
(b) As it is heated the brass expands more than the iron. So the brass forms the outside of a curve with the iron on the inside.
(c) A bimetallic strip can be used in a thermostat to break an electric circuit. When the temperature reaches a certain value the strip bends down enough to open the switch contacts. The heater is then turned off.

How much heat can an object store?

The heat capacity *C* of an object is the heat energy needed to raise its temperature by 1 K.

● Heat capacity *C* is measured in **joules/kelvin (J/K)**. What does the heat capacity of an object depend on? A tank of water for a bath takes much longer to heat up than a kettle of water for a cup of tea. This is because the *mass* of water in the tank is much greater than in the kettle.

● The heat capacity of the object depends on its mass.

An immersion heater is used to heat up a 1 kg block of copper and then 1 kg of water. Both are heated by 10 K (10°C rise). We find that it takes about ten times longer to heat up the water. The heat capacity of water is ten times greater than that of the same mass of copper.

● The heat capacity of an object depends on the material it is made of.

Effects of large heat capacities

The climate of countries, like the British Isles, which are surrounded by sea is affected by the very large heat capacity of the sea water. In winter the sea cools down only slowly. This helps to keep the climate warmer. In the summer the sea is slow to warm up. This keeps the coast cooler. The temperature of the land changes much more quickly than the temperature of the sea.

Cooling the waste water from a large power station is a problem. This is because it holds so much heat energy. Cooling towers are built which allow the warm water to spread out and come into contact with a lot of air. This helps the water to cool. Schemes are being devised to make use of this heat energy which is wasted by power stations.

What is the specific heat capacity of a substance?

The specific heat capacity *c* of a substance is the heat energy needed to raise the temperature of 1 kg of the substance by 1 K (1 degree).

● Specific heat capacity is measured in **J/(kg K)**.
● An *object* has a heat capacity *C* in J/K.
● A *substance* has a *specific* heat capacity *c* in J/(kg K).
● The heat capacity *C* of an object is the specific heat capacity *c* of its material multiplied by its mass *m*: $C = cm$.

The specific heat capacity of water is 4200 J/(kg K). This means that 4200 joules of heat energy are needed to raise the temperature of 1 kg of water by 1 K (1 degree).

A kettle containing 2 kg of water will have a heat capacity $C = cm = 4200 \times 2 = 8400$ J/K.

Calculating the heat energy needed to warm things

$$\frac{\text{heat}}{\text{energy}} = \frac{\text{specific heat}}{\text{capacity}} \times \text{mass} \times \frac{\text{temperature}}{\text{rise}}$$

$$Q = cm\Delta T$$

Example: Calculate the heat energy needed to boil a kettle of water. You are given the following values: mass of water, $m = 2.0$ kg; initial temperature of water = 25°C; final temperature of water = 100°C; specific heat capacity of water, $c = 4200$ J/(kg K):

$$\text{temperature rise, } \Delta T = 100 - 25 = 75$$
$$\text{heat energy, } Q = cm\Delta T = 4200 \times 2.0 \times 75$$
$$= 630\,000\,\text{J}$$
$$\text{or } Q = 630 \text{ kilojoules}$$

38 Change of state

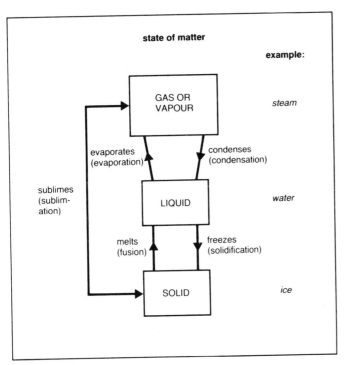

Figure 38.1 *Changes of state:*

- When ice *melts* to form water the change of state is called **fusion**.
- When water turns into vapour at its surface the change of state is called **evaporation**. It can happen at *any* temperature.
- When water turns into steam by forming bubbles within the body of the liquid it boils. The change of state is called **vaporisation**.
- When steam *condenses* to form water the change of state is called **condensation**.
- When water *freezes* to form ice the change of state is called **solidification**.

  When white hoarfrost or snow flakes are formed, water vapour changes directly into ice crystals missing out the liquid state. This change of state is called **sublimation**. Sublimation is a two-way change of state.

Melting and freezing

The temperature at which a solid melts is its melting point. This can be found by plotting a **cooling curve**.

> **A cooling curve is a graph of temperature against time for a sample of a substance as it changes from the liquid to solid state.**

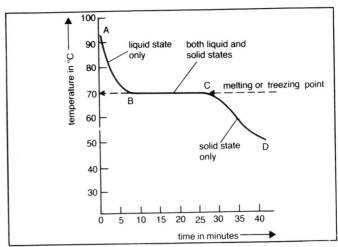

Figure 38.2 *A cooling curve for stearic acid:*

The cooling curve shows a flat section. Here both solid and liquid states exist together. This is the melting or freezing point. During this time heat continues to be lost from the substance as it changes state from liquid to solid but there is *no fall in temperature*.

> **The heat gained when a substance melts and lost when it freezes (without any temperature change) is latent heat of fusion or energy of fusion.**

Effects of latent heat of fusion

- Lumps of ice added to a drink remove heat for a long time. This is because a lot of latent heat is needed to melt the ice.
- Food which contains a lot of ice takes a long time to thaw out when taken from the freezer.
- Larger masses of snow and ice take a long time to melt, even in sunshine. This is because of the latent heat needed to change the state.

What is evaporation?

Evaporation occurs at the *surface* of a liquid. Molecules reaching the surface may either escape or fall back into the liquid. Only the faster molecules actually get away. This is because only they have enough energy to escape from the attraction of all the other liquid molecules. The slower molecules remain in the liquid. So their lower kinetic energy means that the liquid gets *cooler* as it evaporates.

Factors affecting evaporation

The *rate* of evaporation is increased by:

- an increase in the temperature of the liquid (more molecules move fast enough to escape);
- an increase in the surface area of the liquid (more molecules are near the surface);
- a draught or wind blowing over the surface (molecules are carried away so cannot return to liquid).

Some effects of evaporation

- Evaporation of sweat or perspiration from your skin causes a cooling effect.
- When you dry your hands or hair in a stream of hot air it feels cool at first. This is because of the evaporation of water.
- In a cold wind you feel much colder than the air temperature alone would make you feel. The 'chill factor' of the wind is caused by the moving air which increases evaporation from your skin.

Boiling

The temperature at which a liquid boils is called its boiling point.

When a liquid boils, vapour forms bubbles inside the liquid. These rise to its surface and burst. An increase in pressure on the surface of the liquid makes it harder for the bubbles to form. So the boiling point is raised.

- This effect is used in a pressure cooker. The steam pressure inside a pressure cooker is increased. This raises the boiling point of the water. So the food cooks faster at a higher temperature.
- Up a mountain, where the air pressure is lower, water boils at a lower temperature. It can be too low to cook an egg properly.

Vaporising and condensing

Heat is needed to change a liquid into a vapour without any temperature change. This heat is called the latent heat of vaporisation or energy of vaporisation.

An effect of latent heat of vaporisation:

- When steam at 100°C condenses on your skin, it produces a more serious burn than would an equal mass of water at 100°C. This is because the steam also gives up its latent heat of vaporisation.

What is the specific latent heat of a substance?

- *Specific = 'per unit mass'* of a substance.

The specific latent heat of fusion *l* (or specific energy of fusion) of a substance is the heat energy needed to change 1 kg of it from solid to liquid without any temperature change.

- If *l* joules of heat will melt 1 kg of a substance then *ml* joules will melt *m* kg of it.
- Heat energy for melting = *ml* (with no temperature change).

Example: An ice lolly has a mass of 100 g. The specific latent heat of fusion of ice is 340 000 J/kg. Calculate the heat needed to melt the ice lolly. (Assume the ice is at 0°C and no temperature rise occurs.)

$$\text{heat energy for melting} = ml = 0.1\,\text{kg} \times 340\,000\,\text{J/kg}$$
$$\therefore \text{heat needed} = 34\,000 \text{ joules}$$

The specific latent heat of vaporisation *l* (or specific energy of vaporisation) of a substance is the heat energy needed to change 1 kg of it from liquid to vapour without any temperature change.

Example: In chemistry experiment, a salt solution containing 20 g of water at 100°C must be boiled dry. The specific latent heat of vaporisation of water is 2.3×10^6 J/kg. How much heat energy will be needed?

$$\frac{\text{heat energy}}{\text{for vaporising}} = ml = 0.2\,\text{kg} \times 2\,300\,000\,\text{J/kg} = 46\,000\,\text{J}$$

The refrigerator

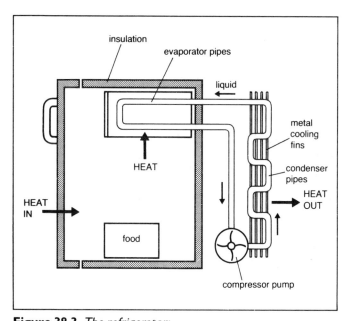

Figure 38.3 *The refrigerator.*

The refrigerator is a **heat pump** which uses the evaporation of a liquid to remove heat from food. The vapour, in a pipe, is pumped out carrying its latent heat with it. The pump *compresses* the vapour back into a liquid in the condenser pipes, causing the vapour to give up its latent heat. This is lost by radiation and convection to the air outside the refrigerator. The liquid then goes back into the evaporator pipes in the freezing compartment. To evaporate again it needs more latent heat. This it removes from the food. The liquid used in the pipes is called Freon. Freon is volatile (evaporates easily).

39
Gases

What is the kinetic theory?

The idea that molecules all have some kind of motion is called the kinetic theory of matter.

● Each moving or vibrating molecule has some kinetic energy.

The total energy possessed by the molecules of an object is called its internal energy.

● When an object is heated it gains internal energy. This is because the kinetic energy of its molecules increases.

● In a gas the molecules move about independently. They have both random direction and random speed.

What evidence is there for the kinetic theory?

Brownian motion

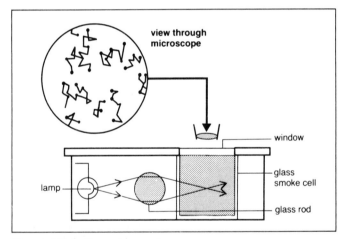

Figure 39.1 *Brownian motion:*

The glass rod acts as a converging lens. This concentrates light in the smoke cell. In the cell we see bright specks. They are dancing about in a jerky, erratic or random way. Molecules are too small to see. They cannot be seen even through a microscope. What we *can* see are the brightly lit smoke particles. They do not often collide with each other. Rather they appear to be knocked about by some other invisible particles. This jerky movement of the smoke particles is known as **Brownian motion**. It is caused by collisions with air molecules. This shows that air molecules move in all directions. They have a range of speeds and kinetic energies.

The same effect can be seen with fine particles suspended in a liquid. This shows that the molecules in a liquid have a similar random motion.

Diffusion

Two kinds of marbles are sorted out in a tray. This is shaken to give the marbles random motion. After a while the marbles become completely mixed up. We could treat the marbles as molecules. Then we have a model of the process called **diffusion**.

Diffusion is the process by which molecules can spread throughout a gas or liquid by their random motion.

Examples of diffusion

● Smells spread to fill the whole of space available. They eventually become too dilute to notice.

● Suppose you add a drop of ink to a glass of water. Its molecules (and their colour) will spread throughout the water without any stirring.

What is Charles' law?

The expansion law for a gas at constant pressure:

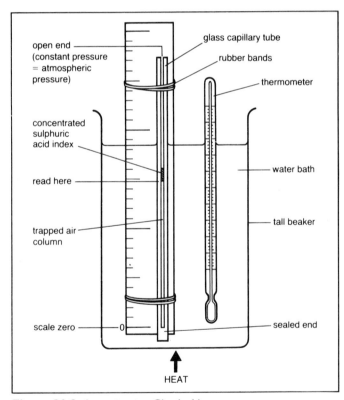

Figure 39.2 *Investigating Charles' law:*

Charles' law, like the other gas laws, applies to a *fixed mass* of gas. The air column in the glass tube is trapped by a drop of acid which moves up the tube as the gas expands. A uniform tube is used so that the length of the column can be taken as a measure of the volume of air trapped. As the water bath is heated, readings are taken of the temperature and the length of the air column.

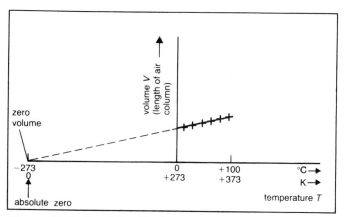

Figure 39.3 *Graph for Charles' law.*

This is a volume against temperature graph. It gives a straight line. This demonstrates **Charles' law**:

> **The volume *V* of a fixed mass of gas is directly proportional to its absolute temperature *T* (on the Kelvin scale) if the pressure is constant.**

$$\frac{V}{T} = \text{constant} \qquad \frac{V_1}{T_1} = \frac{V_2}{T_2}$$

Absolute zero

The graph also predicts that at $-273\,°C$, absolute zero, the volume of the gas should contract to zero. It is the lowest possible temperature. Here the molecules have their minimum energy and no motion.

What is Boyle's law?

Compressing a gas at constant temperature

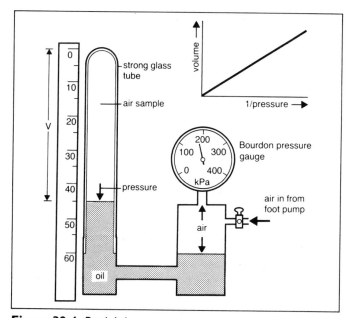

Figure 39.4 *Boyle's law.*

Oil traps a fixed mass of air in the strong glass tube. A car foot-pump is used to increase the pressure. This is read on the Bourdon gauge. The volume of the trapped air is shown by the length of the air column and measured on the cm^3 scale.

A graph is plotted of volume against 1/pressure. It gives a straight line through the origin. This shows **Boyle's law**:

> **The volume *V* of a fixed mass of gas is inversely proportional to its pressure *p* if the temperature is constant.**

$$pV = \text{constant} \qquad p_1V_1 = p_2V_2$$

What is the pressure law?

Heating a gas at constant volume

If a gas is heated but not allowed to expand its pressure rises. A graph is plotted of pressure against temperature. It is similar to the one obtained for Charles' law. It gives the **pressure law**:

> **The pressure *p* of a fixed mass of gas is directly proportional to its absolute temperature *T* if its volume is constant.**

$$\frac{p}{T} = \text{constant} \qquad \frac{p_1}{T_1} = \frac{p_2}{T_2}$$

How do we use the gas equation?

Combining the three gas laws gives the **gas equation**:

$$\frac{pV}{T} = \text{constant} \qquad \frac{p_1V_1}{T_1} = \frac{p_2V_2}{T_2}$$

Rules for using this formula:

- The mass of gas must be constant.
- The temperatures T_1 and T_2 must be given in kelvins
- The units in which p and V are calculated must be the same on both sides of the equation.

Example: A bicycle pump holds $60\,cm^3$ of air (V_1) when the piston is drawn out. The air is initially at $17\,°C$ and 1.0 atmospheres pressure (p_1). Compression reduces its volume to $15\,cm^3$ (V_2). The temperature is raised to $27\,°C$. Calculate the pressure (p_2) of the air as it is forced into the tyre.

Temperatures must be calculated in kelvin:

$$T_1 = 17 + 273 = 290\,K, \quad T_2 = 27 + 273 = 300\,K$$

Rearranging the formula gives:

$$p_2 = p_1 \times \frac{V_1}{V_2} \times \frac{T_2}{T_1} = 1.0 \times \frac{60}{15} \times \frac{300}{290} = 4.1 \text{ atmospheres}$$

40

Heat energy gets around

Conduction

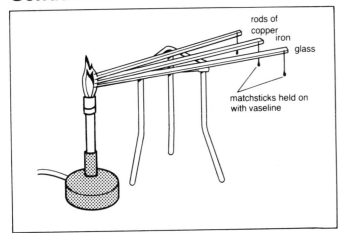

Figure 40.1 *Comparing rates of heat conduction in different materials:*

Heat flows along the rods at different rates. The matchsticks drop off in the order: copper, iron, glass.

The flow of heat through a material without the material itself moving is called conduction.

Metals are good conductors of heat.

Some solid materials such as glass, wood and plastic conduct heat only slowly. They are called insulators.

How does conduction work?

Conduction is a flow of heat. This happens when energy is passed from molecule to molecule. The energy we call heat is the kinetic energy of vibration of molecules. Each molecule can pass on some of its kinetic energy by 'bumping' into its neighbours and making them vibrate more. So energy is passed from molecule to molecule. This is the mechanism of conduction in all materials including liquids and gases.

This conduction process usually works slowly in solids and liquids. It is very slow in gases, where the molecules are generally further apart.

However, a metal contains free electrons. These can move independently through the metal. When a metal is heated, these free electrons move faster with more kinetic energy. The electrons spread by diffusion into cooler parts of the metal. They collide with the molecules, transferring their kinetic energy. Heat flows very quickly through metals by this mechanism.

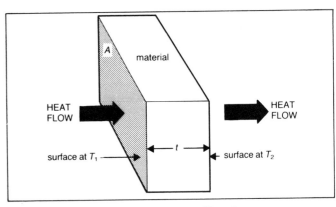

Figure 40.2 *What affects the flow of heat?*

The flow of heat is affected by:

- the nature of the material (insulator ?);
- the surface area, A, through which heat can flow – the bigger the area the faster the heat flows (the area of a window for example);
- the thickness of the material, t – the thicker the material the slower the heat flows;
- the temperature difference $(T_1 - T_2)$ across the thickness – the greater the temperature difference the faster the heat flows.

How can buildings be insulated?

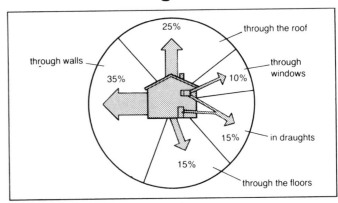

Figure 40.3 *Heat losses from a house:*

Insulation can be used to save heat losses from a house:

- 75 mm of mineral or glass-fibre insulation can be used in the loft. It can save 80% of heat losses through the roof.
- Mineral fibre or foam insulation can be put in the cavity (space) between the outer walls. It can save 65% of heat losses through the walls.
- Double glazing of windows can save 50% of heat losses through the windows. But this will be only about 5% of the total losses from a house. Insulating shutters can be fitted to windows. This reduces the heat losses by as much as 80%.
- Thick carpets, curtains and draught-excluders can save much more heat than double glazing.

Convection

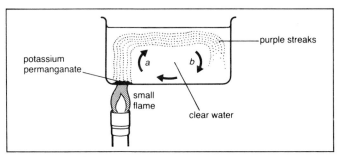

Figure 40.4 *Convection currents in water:*

A few small crystals of potassium permanganate are placed in the water at one end. The dish underneath them is heated. Purple streaks can be seen rising with the water above the crystals (a). At the far side, away from the heat, the water sinks (b). The whole body of the water circulates by what is called a **convection current.**

● The convection current carries heat with it. The hot water rises because when heated it expands and becomes less dense. So it floats above colder and denser water. Cold water sinks because it is more dense than surrounding water.

● Convection currents also happen in gases. The apparatus below demonstrates convection in air.

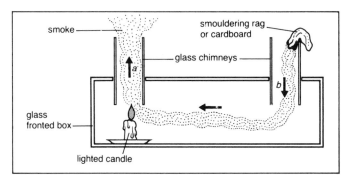

Figure 40.5 *Convection currents in air:*

Natural convection in a hot-water system

Natural convection is often used to carry hot water up from the boiler to the hot-water storage tank. This does not need a pump. Hot water flows into the top of the storage tank. At the same time cold water flows down through a pipe from the bottom of the tank to the boiler.

Radiation

Heat can also travel as radiation. It does so as **infra-red (IR) radiation.** IR is part of the electromagnetic spectrum. So it moves at the speed of light. It can travel in a vacuum.

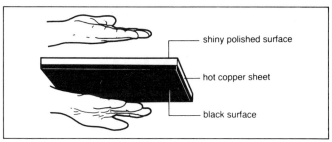

Figure 40.6 *Emission of radiant heat:*

The side painted matt black feels much hotter than the shiny polished side.

● A *dull black* surface is both a good emitter of IR and a good absorber.
● A *light-coloured* or *shiny surface* is both a poor emitter of IR and a poor absorber.

How can we detect IR radiation?

IR radiation can be detected a number of ways:

● **By hands.** This is as in the demonstration above.
● **By thermopile.** This is a bank of cells which convert heat energy into a small thermoelectric voltage.
● **By phototransistor.**
● **By photography.** Thermal imaging uses either IR-sensitive photographic paper or electronic scanning systems which respond to IR.

Remember that IR is invisible. The red glow we see in a hot fire is visible light. It is not IR.

How does the vacuum flask work?

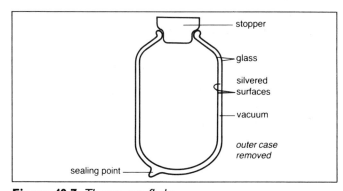

Figure 40.7 *The vacuum flask:*

● Conduction and convection are stopped by the vacuum between the double glass walls. (The sealing point is where air was sucked out.)
● The cork or plastic stopper contains trapped air. This is a good insulator.
● Heat loss by infra-red radiation is reduced by the two silver coatings on the glass walls of the flask. These are both bad emitters of radiation and good reflectors. This keeps the radiation inside the flask.

Review questions: Chapters 37, 38 39 and 40

1 Copy and complete the following:
   (a) Heat is a form of _____.
   (b) A measure of 'hotness' or 'coldness' of an object is called its _____.
   (c) Temperature is measured with a _____.

2 On the Celsius scale what are the names and values of the: (a) upper; (b) lower fixed points?

3 (a) What is the name of the lowest temperature which can exist?
   (b) What is the value of the lowest possible temperature in: (i) kelvin; (ii) °C?

4 (a) Copper melts at 1356 K. What is this in °C?
   (b) Helium boils at −269 °C. What is this in kelvin?

5 What is the name of a thermometer which:
   (a) is used to measure body temperature;
   (b) changes its resistance with temperature;
   (c) consists of two metals joined together;
   (d) is used in very cold places?

6 (a) What happens to the kinetic energy of molecules when they are heated?
   (b) What happens to the size of materials when they get hotter?
   (c) Why does a bimetallic strip have two different metals riveted together?

7 Copy and complete the following:
   (a) Heat _____ is the energy needed to raise an object's temperature by 1 K.
   (b) _____ heat capacity is the energy needed to raise the temperature of 1 kg of a substance by 1 K.

8 The specific heat capacity of water is 4200 J/(kg K). What is the heat capacity of:
   (a) 1 kg of water; (b) 5 kg of water?

9 Two men wish to make a pot of tea. The kettle holds 1.5 kg of water which is at 20 °C. What is:
   (a) the final temperature of the pure water when it boils (in °C);
   (b) the heat energy needed to boil the water if c for water is 4200 J/(kg K)?

10 What changes of state occur in the following:
   (a) vaporisation;
   (b) condensation;
   (c) solidification;
   (d) fusion;
   (e) sublimation?

11 A girl carries out a cooling curve experiment. She plots a graph. This is shown in the figure.

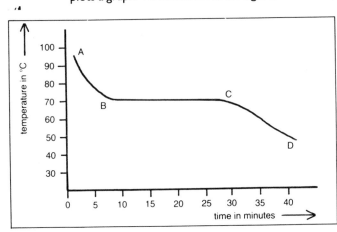

   (a) Which part of the graph shows where both solid and liquid states exist together?
   (b) What is the melting point of this substance?
   (c) What is the name of the heat lost while the substance changed state?

12 (a) What happens to the temperature while a substance changes state?
   (b) The specific latent heat of fusion of gold is 70 000 J/kg. Calculate the heat needed to melt 5 g of gold.

13 A girl holds a mug of hot chocolate.
   (a) Where does evaporation take place?
   (b) Do the faster or slower molecules leave the liquid?
   (c) What happens to the temperature of her drink?
   (d) Why might she 'blow' over the top of the mug?

14 (a) What happens in a liquid when it boils?
   (b) What happens to the temperature while the boiling liquid changes into a vapour?
   (c) The specific latent heat of vaporisation of water is 2.3×10^6 J/kg. Calculate the heat needed to change 2 kg of water at 100 °C into steam at 100 °C.

15 (a) What is the name given to the idea that molecules all have some motion?
   (b) What type of energy do the molecules have due to their motion?
   (c) What is the total energy of the molecules of a material called?

16 In a smoke cell experiment:
  (a) What are the bright specks moving about?
  (b) Describe how these specks move about.
  (c) Why do these specks move about?
  (d) What is the name of this effect?

17 The figure shows two liquids poured into a jar to form separate layers.

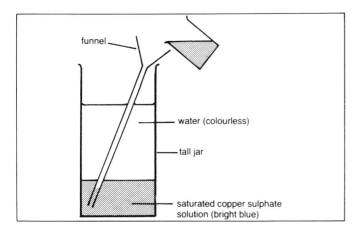

funnel

water (colourless)

tall jar

saturated copper sulphate solution (bright blue)

  (a) Describe how the molecules are moving.
  (b) What will the colour of the liquids be after a few days?
  (c) What is the name of this process?

18 A fixed mass of gas is kept at a constant pressure.
  (a) What happens to the volume of the gas if the temperature in kelvin:
    (i) doubles; (ii) increases by three times?
  (b) The volume of the gas is 30 litres at 200 K. What is the volume at 300 K?

19 A fixed mass of gas is kept at a constant temperature:
  (a) What happens to the volume if the pressure:
    (i) doubles; (ii) halves?
  (b) The volume of the gas is $0.4\,m^3$ at $4\,kN/m^2$. What is the volume at $20\,kN/m^2$?

20 A fixed mass of gas is kept at a constant volume:
  (a) What happens to the pressure if the temperature in kelvin:
    (i) doubles; (ii) reduces by three times?
  (b) The pressure of the gas is 75 kPa at 200 K. What is the pressure at 400 K?

21 A gas sample has a volume of $3\,m^3$. Its temperature is 27 °C. The pressure is 1 atmosphere. Find its volume at:
  (a) 127 °C and 0.5 atmospheres.
  (b) −73 °C and 2 atmospheres.

C40 **22** (a) What is the name of the flow of heat through a material without the material moving as a whole?
  (b) How is the heat energy moved through the material?
  (c) Why does this process work only very slowly in gases?
  (d) Why does this process work fastest in metals?

23 A lady wishes to fit a new back door which will help to reduce the rate of heat loss from her kitchen. Should she fit:
  (a) a wood or aluminium door;
  (b) a single or double width door;
  (c) a thin or thicker door?

24 What is the name of the flow of heat through a liquid or gas where the molecules move from one place to another?

25 A saucepan with water is placed onto a hotplate.
  (a) What is the name of the method that transfers heat through the saucepan to the water?
  (b) Describe how convection currents are set up in the water.

26 (a) What is heat radiation called?
  (b) What is the speed of heat radiation?
  (c) What type of surfaces are good absorbers of heat radiation?
  (d) What type of surfaces are bad emitters of heat radiation?

27 In a vacuum flask:
  (a) What are the walls made of?
  (b) What are the walls coated with?
  (c) What is between the walls?
  (d) What is the stopper made of?
  (e) Which of the above are intended to reduce heat loss by conduction?

People need energy

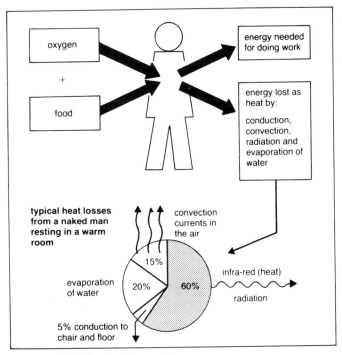

Figure 41.1 *Our bodies need energy:*

Our energy input is 10 to 20 MJ per day. This amount of food is needed by our bodies:

- to grow;
- to repair and replace cells;
- to keep us warm because we lose heat;
- to drive the electrical system used by our brains and nerves;
- to combine with oxygen to provide the energy we need to move, work and live.

Some energy is stored as fat or held ready for use in muscles. Energy is released from food when it combines with oxygen.

The loss of heat from our bodies

This depends on:

- the surrounding air temperature;
- whether there is a breeze to aid evaporation;
- what clothes we are wearing;
- the surface area of our bodies;
- the kind of activity we are doing.

When people are working very hard or perhaps running in a race, they then need to lose much more heat. Sweating helps them do this. When sweat evaporates taking its latent heat from the body, cooling is increased by a factor of 10.

Activity of a 70 kg person	Rate of working in joules/second or watts
Sleeping	80
Sitting reading	120
Typewriting or piano playing	160
Walking slowly	250
Running, swimming, hard work	500–800
Walking up stairs	1300

How does energy damage our bodies?

- If you put your hands into a fire you get burnt. Heat energy is received too quickly.
- A bullet fired at high speed by a gun has a lot of kinetic energy. This makes it dangerous. It is the energy which does the damage.
- A fast-moving car and its occupants has kinetic energy. It is this that causes damage and injury in a collision.
- Alpha particles can damage body cells. This is because they have energy.

Do machines need energy?

Suppose we want a machine to do work for us. It must have a supply of energy. Energy is conserved. So the work output of a machine can never exceed the energy input.

In the industrial revolution, in the late 18th century people began to use fire to release the energy stored in fossil fuels like coal. Now we get energy from many sources. Examples are oil, gas, wind, the Sun and the nuclei of atoms. People who can supply energy to machines when and where they need it have much more control over the world around them. Such people are freed by machines from much of the hard work they would otherwise have to do themselves.

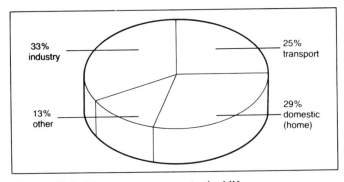

Figure 41.2 *How we use energy in the UK:*

Over half of the energy is used by industry. This is because most of the energy used for transport is also helping industry. The supply of energy is vital for our industrial society.

How do we use energy at home?

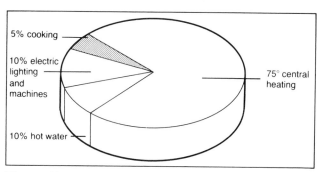

Figure 41.3 *How we use energy in the home:*

- Most of the energy used at home is needed to keep us warm or to heat water.
- 10% of the energy powers our home machines.

We have machines which wash clothes and dishes, mix food, clean floors, polish, drill and sew for us. While some of these machines need our attention, they all take away most of the human effort by using another source of energy to do the hard work.

Calculating the energy used at home

The table below can be used to convert the various energy units used for the different kinds of fuel into megajoules (MJ).

Energy source	Unit in which you buy it	Equivalent in MJ
Electricity	Kilowatt-hour	3.6
Gas	Therm	110
Coal	Tonne (= 1000 kg)	28 000
Fuel oil	Litre	38

Example: *Electricity bills.*

present reading	54892
previous reading	52642
units used:	02250

These 'units' are kilowatt-hours = 3.6 MJ:

$$\therefore \text{energy used} = 2250 \times 3.6\,\text{MJ} = 8100\,\text{MJ}$$

Example: *Gas bills.* The gas used is first measured in hundreds of cubic feet of gas. This is converted on the gas bill into therms. Each therm = 110 MJ:

$$\therefore 400 \text{ therms would be } 400 \times 110\,\text{MJ} = 44\,000\,\text{MJ}$$

The total energy used in a house in the UK can vary over a wide range. But a family using central heating can easily use 10^{11} joules or 100 000 MJ per year.

The price of different forms of energy

Energy source	Unit	Price per unit in £
Electricity	Kilowatt-hour	0.055
Gas	Therm	0.40
Coal	Tonne	120
Fuel oil	Litre	0.20

Example: *Buying electricity.* £0.055 (5½p) buys 1 unit of electricity or 3.6 MJ:

$$\therefore \text{£1 will buy } \frac{3.6\,\text{MJ}}{0.055} = 65\,\text{MJ of energy}$$

Example: *Buying coal.* £120 buys 1 tonne of coal or 28 000 MJ:

$$\therefore \text{£1 will buy } \frac{28\,000\,\text{MJ}}{120} = 233\,\text{MJ of energy}$$

- Electricity is much more expensive than the other energy sources.

How can we save energy at home?

Most energy saving can be made by cutting down the amount of heat needed to keep a house warm. The rate of loss of heat from a house can be reduced by:
- having a lower temperature inside the house;
- increasing the insulation around the house.

Other energy saving ideas are:

1. Lag (wrap with insulating material) hot-water pipes and tanks.
2. Turn down the temperature setting on the hot-water thermostat.
3. Block off unused fireplaces. This stops heat going up the chimney.
4. Fit linings to curtains. Always close them at night.
5. Do not heat rooms that are not used. Keep their doors shut.
6. Fit separate thermostats to the radiators in each room. Some rooms can then be cooler.
7. Do not wash up in running hot water.
8. Use low-temperature washing powders and programmes for washing machines.
9. Use a washing line and free wind and sunshine instead of tumble-dryer whenever possible.
10. Do not put warm food into a fridge or freezer.
11. Switch off all lights which are not needed. Fit low-power or fluorescent lighting where it is acceptable.
12. Do not put more water in a kettle than you need for your hot drinks (as long as the element is covered). Do not leave it boiling.

42
Energy sources

Sources of energy.

	Source	Type of energy	Uses as an energy source
From the Sun today 1.7×10^{17} watts	* Sunlight * Radiant heat * Wind power * Wave power * Water power * Food plants, wood * Biogas (methane)	Radiation Radiation Mechanical kinetic Mechanical kinetic Gravitational potential Biological or 'biomass'	Electricity from solar cells Solar panels, solar furnace Windmills to generate electricity Generators to provide electricity Hydroelectric power: electricity Food for people and animals Fuel
From the Sun going back 600 million years	Coal Oil Natural gas	Chemical	Conventional power stations to produce electricity, transport and heating
From the Earth	* Geothermal * Tidal energy Nuclear power Nuclear power	Heat Gravitational potential Fission Fusion	Local heating Tidal barrage: electricity Nuclear power stations: electricity ?Power station of the future: electricity

* Renewable sources

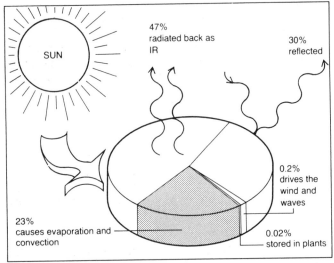

Figure 42.1 *What happens to the Sun's energy arriving at the Earth?*

The Sun supplies 5000 times more energy than all other sources. Over 75% of the Sun's energy arriving at the Earth is quickly lost. The Sun can supply us with energy *directly*:

- Solar cells turn light into electricity.
- Solar panels absorb IR radiation to heat water.

The Sun supplies **renewable** energy:

- by driving the winds and waves;
- by lifting water vapour into the clouds which fill rivers and reservoirs with water to drive hydroelectric schemes;
- by growing food and producing 'biomass'.

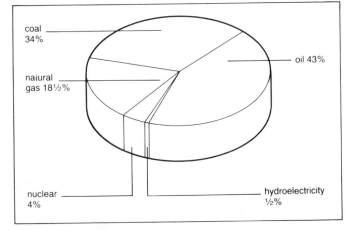

Figure 42.2 *UK primary energy sources:*

Primary sources of energy are those forms in which the energy is first supplied. So, for example, coal is a primary source but electricity generated by burning coal is not. Over 95% of our primary sources are the *non-renewable* fossil fuels. These are oil, coal and gas. Over the last 200 years we have been rapidly using up the non-renewable sources of energy.

What alternative sources of energy are available?

Questions to ask about alternative sources are:

1. Is the source renewable?
2. How much power could be supplied by this method?
3. How far developed is the necessary technology?

④ Are there any risks to the environment?

⑤ What would the scheme cost to develop and build?

⑥ What would be the energy price?

For comparison, the estimated cost of energy from a new nuclear power station is 3p per kWh. From a new coal-fired station it is 4.3p per kW h.

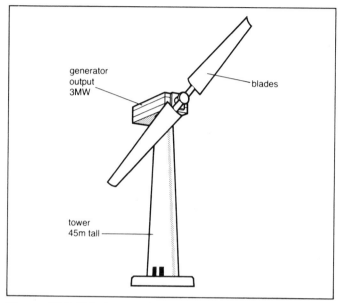

Figure 42.3 *Wind energy:*

This 60-metre-diameter aerogenerator, built on the Orkney Islands, can produce up to 3MW of power. The energy source is renewable and 'free'. The technology is already being tested. There are no environmental risks. However, suitable windy sites may also be areas of natural beauty. Rotating metal blades may interfere with TV reception. Wind 'farms' are being planned. In these many aerogenerators could produce a power output similar to that of a conventional or nuclear power station. The cost per kilowatt hour is 2.5 to 3.2p.

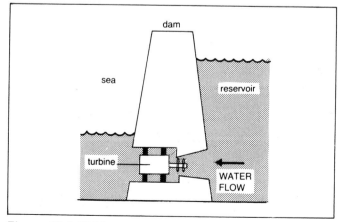

Figure 42.4 *Tidal energy:*

A tidal barrage across the Bristol Channel has been considered for several years. The incoming tides raise water to a high level behind a dam. As the tide goes out, the trapped water is released through holes in the dam. There it drives turbines. Up to 5000MW of power could be generated. The scheme would be very expensive to build. However, tidal energy is renewable and 'free'. The cost per kilowatt hour is 3 to 7p.

Geothermal energy

Rocks below the ground get warmer as you go deeper into the Earth. Heat can be extracted by sending cold water down and pumping hot water or steam back up again. Holes have been bored down 12 km in the granite rocks in Cornish mines. They reach temperatures of 200 to 300°C. Drilling boreholes beneath cities would be very noisy. Undesirable gases may escape from underground. These include radioactive radon. The cost per kW h is 3 to 6p.

Wave energy

In the Atlantic, waves carry about 80 kW of power per metre of wave frontage. Scale models of a rocking-boom type of wave-driven generator have been built. They are capable of converting 50% of this energy to electricity. However, building and maintaining full scale generators would be difficult. So wave power can only be a hopeful long shot for the future. The cost per kW h is 9 to 15p.

Solar energy

Solar cells convert solar energy directly into electricity. These are already used in watches and calculators and to power satellites. They are unlikely to be used to generate large quantities of electrical power in the near future. This is because:

● the solar cells are very expensive;
● on average only about 10 watts of power can be obtained from a square metre of cells.

Solar panels use solar radiant energy (IR) to provide hot water. They are much more successful. A transparent cover traps the solar radiation in the manner of a green-house. A black surface helps absorb the radiation.

Heat pumps

Heat pumps are not a source of energy in themselves. A refrigerator is a heat pump. Heat pumps can be used to take heat from the air outside a house. They add it to the air inside the house. To do this the condenser which releases heat is inside. The evaporator which absorbs heat is outside. The electrical energy needed to drive the pump is several times smaller than the amount of heat energy pumped into the house.

Energy conversions

Form of energy	Where or how it is found	How it is used or what it does
Stored	• In chemicals • As elastic potential energy • As gravitational potential energy • In the nuclei of atoms	Food; fuel; batteries and rockets Catapult; bow and arrow; compressed or wound up springs; springboard Objects at a height h have stored energy given by: $E_p = mgh$ Nuclear fission: nuclear power stations and in the atom bomb
Kinetic	• In moving objects • As heat energy of moving molecules • As sound energy	Can be calculated using $E_k = \frac{1}{2}mv^2$ Causes conduction and convection of heat, diffusion and evaporation Vibrations of molecules transmit sound waves through materials
Electromagnetic radiation	As radio waves, IR, light, UV, X-rays and γ-rays	Electromagnetic radiation is pure energy travelling at 3×10^8 m/s; (see chapter 22 for uses)
Electricity	• In nature • Produced by generators or batteries	Lightning; our brains communicate with the rest of our bodies via the nerve cells using electricity Electricity is energy 'on the move'; essential for electric motors; for electronic circuits and computers

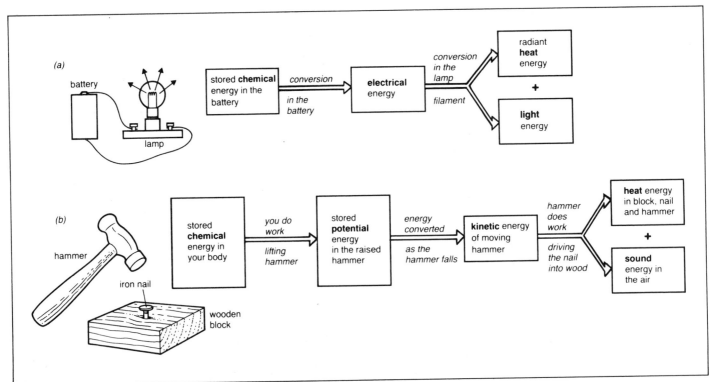

Figure 43.1 *Energy conversions: (a) A battery and lamp. (b) A hammer and nail:*

The *blocks* contain the different forms of energy. The *arrows* show the energy changes or conversions. The linked forms of energy make an '**energy chain**'.

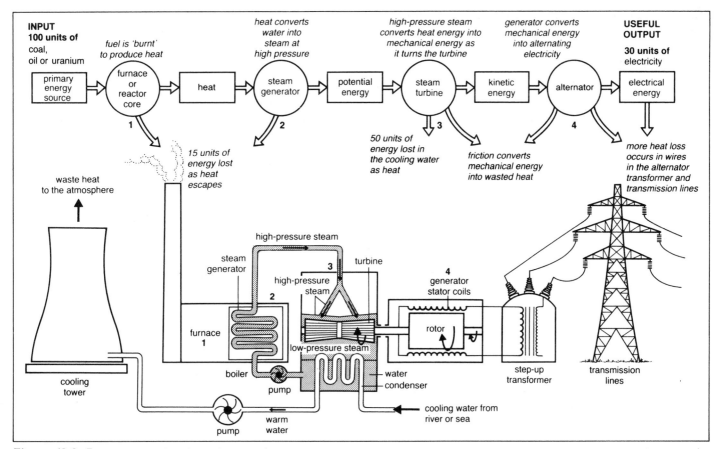

Figure 43.2 *Energy conversions in a power station.*

What is a pumped storage scheme?

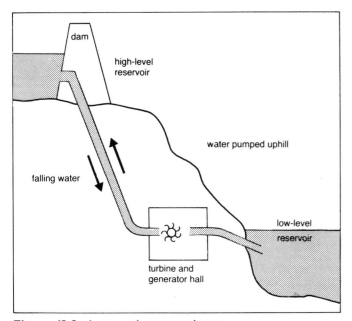

Figure 43.3 *A pumped storage scheme:*

A pumped storage scheme is a very successful method of saving spare electrical energy when the demand for it is low. The spare electrical energy is used to pump water from a low-level reservoir to a high-level one. The energy is stored as gravitational potential energy in the high-level water.

When the demand for electrical energy rises, water is released from the upper reservoir. The falling water gives up its stored potential energy to turn turbines. These turn electricity generators.

The Dinorwig scheme in North Wales can deliver 1800 MW of electrical power continuously for 5 hours from a full reservoir. Equally important, it can reach 1300 MW of output power within 10 seconds of a call for electricity. For every four units of electrical energy used to pump water uphill, the scheme returns three units of electrical energy after storage.
How efficient is the scheme?
Where has the other unit of energy gone?

Energy conservation

No energy is ever destroyed in all the energy changes we can find. We say that energy is **conserved**. However, energy often appears to be 'lost'. Energy which starts as a high-grade stored energy or as electrical energy tends to end up as waste heat energy.

Review questions: Chapters 41, 42 and 43

C41

1 Copy and complete the following:
   (a) Food provides materials for repairing and replacing our body _____.
   (b) Food combining with _____ releases energy.

2 The figure shows typical heat losses from a man resting on a chair in a warm room.

A: 60% IR radiation
B: 5% conduction to chair and floor
C: 20% evaporation of water
D: 15% convection currents in air

   (a) How is most heat lost?
   (b) How is least heat lost?
   Which type of heat loss would increase if:
   (c) he did some weight lifting so that he was sweating?
   (d) an electric fan was switched on?
   (e) he laid stretched out on the floor?

3 (a) What is the main use of energy in the UK?
   (b) What percentage of energy is used for transport?
   (c) What is most energy in our homes used for?
   (d) What percentage of our energy is used for cooking?

4 What units are used on the following bills:
   (a) electricity; (b) gas; (c) coal; (d) oil?

5 Beryl's electricity meter reads 64 482. The previous reading was 61 982.
   (a) How many units has she used?
   (b) Each unit is worth 3.6 MJ. Find how much energy she has used.

6 Tariq has used 300 therms of gas. Each therm is 110 MJ of energy. How much energy has he used?

7 Which of the following would help to save energy at home?
   (a) opening curtains at night;
   (b) fitting linings to curtains;
   (c) washing up in running hot water;
   (d) insulating hot-water pipes and tanks;
   (e) using high-temperature washing powders;
   (f) using fluorescent lights where possible;
   (g) putting warm food in the fridge?

8 Use the data in the table below to answer the questions:

Energy source	Equivalent energy in MJ	Unit price in £	£1 will buy
1 kWh of electricity	3.6	0.055	65 MJ
1 therm of gas	110	0.40	
1 tonne of coal	28 000	120	233 MJ
1 litre of oil	38	0.20	

   (a) For each energy unit, which source:
       (i) gives the largest energy in MJ;
       (ii) gives the smallest energy in MJ;
       (iii) has the most expensive price;
       (iv) has the cheapest price?
   (b) Calculate what £1 will buy in MJ of:
       (i) gas energy; (ii) oil energy.
   (c) Which source gives the best value for money?

C42

9 Which energy source is used by:
   (a) windmills;   (b) solar cells;
   (c) tidal barrage;   (d) solar panels;
   (e) conventional power stations?

10 (a) What happens to most of the Sun's energy arriving at the Earth?
   (b) What percentage of the energy arriving at the Earth drives the wind and waves:
       (i) 20; (ii) 2; (iii) 0.2; (iv) 0.02?

11 Which of the following are *primary* sources of energy: (a) coal; (b) electricity;
   (c) oil; (d) nuclear; (e) heat pumps?

12 Which of the following energy sources are *renewable*:
   (a) coal; (b) wind; (c) natural gas; (d) oil;
   (e) tidal; (f) water wave; (g) radiant heat?

13 (a) What is the energy source used by aerogenerators?
   (b) What *type* of energy is this source?
   (c) What is a typical length of the blades (tip to tip): (i) 0.06 m; (ii) 0.6 m;
       (iii) 6 m; (iv) 60 m; (v) 600 m?

14 (a) Describe the main features of a tidal barrage.
   (b) What *type* of energy is tidal power?

15 (a) What is geothermal heat?
   (b) How is this heat extracted?
   (c) What else may be released from the ground?

16 The figure shows the cross-section of a rocking-boom wave-driven generator.

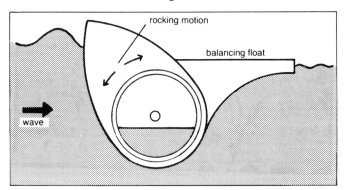

Out in the Atlantic, waves carry about 80 kW of power per metre of wave frontage.
   (a) How much power would be carried in 25 km of wave frontage?
   (b) Suppose the wave generator shown was built 25 km long. What is the power output if it converts only 50% of the wave power?
   (c) Explain why the waves are much smaller after they have passed the generator.

17 (a) Why are solar cells unlikely to be used to generate large quantities of electricity in the foreseeable future?
   (b) Why are solar panels painted black?

18 (a) What do heat pumps transfer?
   (b) What is the purpose of the:
     (i) condenser; (ii) evaporator?
   (c) Which should be on the inside of a refrigerator, the condenser or the evaporator?
   (d) When a pump is used for heating, should its condenser be inside or outside the house?

C43 19 What type of energy is associated with:
   (a) a dish of strawberries; (b) lightning;
   (c) a mouse running; (d) an inflated balloon;
   (e) a girl shouting; (f) water in a reservoir?

20 Name a device which can convert:
   (a) sound energy into electrical energy;
   (b) chemical energy into electrical energy;
   (c) electrical energy into light energy;
   (d) electrical energy into kinetic energy;

21 Energy chains can be drawn with arrows and blocks.
   (a) What do the arrows show?
   (b) What do the blocks contain?
   (c) What is an energy chain called which can go in both directions?

22 Describe the energy changes for a lighted candle.

23 The figure shows a block diagram for a battery-driven electric motor raising a load.

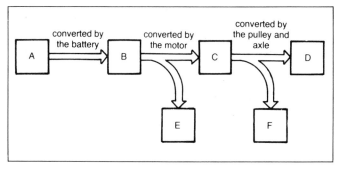

   (a) What type of energy is shown by blocks A, B, C and D?
   (b) What are the causes of the heat losses shown by blocks E and F?

24 In a power station:
   (a) Name a *primary* energy source.
   (b) What are the main forms of energy in the chain?
   (c) What *type* of energy is lost?

25 (a) Why is a pumped storage scheme useful?
   (b) When the water is pumped, does it flow uphill or downhill?
   (c) In what form has the surplus electrical energy been stored?

26 A hydro-electric power station uses a reservoir with a water level 40 m above the turbines. The mass of water flowing through the turbines every second is 5000 kg/s.
   (a) What is the potential energy of 5000 kg of water at a height of 40 m?
     (Assume $g = 10$ N/kg.)
   (b) How much energy/second does the water deliver to the turbines?
   (c) If the efficiency of the station is 80%, what is the power output?

27 (a) What is meant by *conservation of energy*?
   (b) What has energy often become when it seems to be 'lost'?

44
Electricity distribution

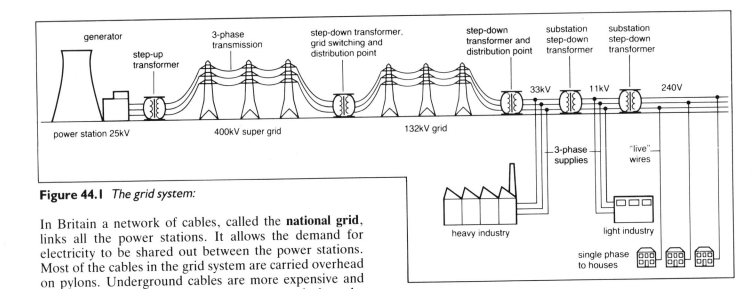

Figure 44.1 *The grid system:*

In Britain a network of cables, called the **national grid**, links all the power stations. It allows the demand for electricity to be shared out between the power stations. Most of the cables in the grid system are carried overhead on pylons. Underground cables are more expensive and difficult to maintain. They are used in cities and where the scenery must not be spoilt.

How much power is lost in cables?

For a cable of resistance R, carrying a current I, the undesirable effects of power losses are:

- a power loss in the cable as heat $= I^2 R$;
- a drop in voltage along the cables $= IR$:

$$V_{out} = V_{in} - IR$$

Power loss in cables can be reduced if:

- thick wires with a low resistance are used.
- power is transmitted at high voltages. At higher voltages the same power can be transmitted at lower current.   power = current × voltage

Example (i): $10\,MW = 400\,A \times 25\,kV$
Transmission of 10 MW of power at 25 kilovolts requires a high current of 400 amps.

Example (ii): $10\,MW = 25\,A \times 400\,kV$
By increasing the voltage to 400 kV, as in the super-grid, the current can be made 16 times smaller. This reduces the power loss in the cables by 16^2 times = 256 times smaller!

Calculating the power loss

If the cable in the examples above had a total resistance of $5\,\Omega$, the power loss would be:
Case (i): power loss $= I^2 R = 400^2 \times 5 = 0.8\,MW$.
Case (ii): power loss $= 25^2 \times 5 = 3125$ watts.

Calculating the voltage drop

Case (i): voltage drop along cable $= IR$

$$IR = 400 \times 5 = 2000\,V \text{ or } 2\,kV$$
$$\therefore V_{out} = V_{in} - IR = 25\,kV - 2\,kV = 23\,kV$$

This is a serious 8% drop in voltage.
Case (ii): $IR = 25 \times 5 = 125$ volts.
A voltage drop of 125 volts is very small compared with 400 kilovolts!

What are transformers used for?

Transformers can be used to *step-up* or *step-down* a voltage. In the national grid both are used so that the power can be transmitted at high voltages but used at safer low voltages.

$$\frac{V_{out}}{V_{in}} = \frac{n_s}{n_p} \qquad V_{out} = V_{in} \times \frac{n_s}{n_p}$$

If the step-up transformer at the power station has 500 turns on its **primary coil** and 8000 turns on its **secondary coil**, when the input voltage is 25 kilovolts, the output voltage will be:

$$V_{out} = V_{in} \times \frac{n_s}{n_p} = 25\,000 \times \frac{8000}{500} = 400\,000\,V$$

The structure of atoms

The Rutherford–Bohr model of the atom

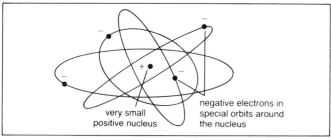

Figure 45.1 *The Rutherford–Bohr model:*

In about 1913 Niels Bohr suggested that the atom had a small central **nucleus** with electrons in orbit around it like planets around the Sun. The evidence for the small *positively charged* nucleus came in 1906. This was from the α-particle-scattering experiments of Ernest Rutherford.

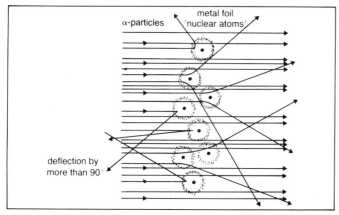

Figure 45.2 *α-particle scattering:*

Rutherford fired a stream of alpha particles at a thin metal foil. He found that:

- most of the α-particles passes straight through;
- a very small fraction (about 1 in 8000) were deflected by more than 90°. That is, they bounced back towards the source. Rutherford was very surprised by this. He said: 'It was about as credible as if you had fired a 15 inch shell at a piece of tissue paper and it came back and hit you.'
- The nucleus of an atom is very small. It is only about $\frac{1}{10000}$ of the diameter of the whole atom. Most of the atom is empty space. This allows the very small α-particles to pass straight through.

- The nucleus is *positively charged*. It repels the positive α-particle as it approaches close.

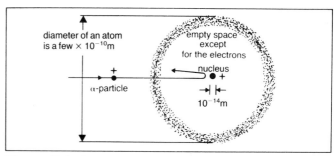

Figure 45.3 *Comparing atomic and nuclear sizes:*

The scale of the nucleus is tiny compared with its atom. It can be likened to a walnut at the centre of Wembley Stadium, or a pin-head inside a house.

Particles in the atom.

Particle	Symbol	Charge	Relative mass	Place in the atom
Electron	e	$-e$	(m_e)	Orbits round the nucleus
Proton	p	$+e$	$1836 m_e$	**Nucleons**, i.e. particles in the nucleus
Neutron	n	Neutral	$1839 m_e$	

The charge of one electron $-e = -1.6 \times 10^{-19}$ coulombs. The mass of one electron $m_e = 9.1 \times 10^{-31}$ kg.

What are isotopes and nuclides?

The proton number or atomic number Z = the number of protons in the nucleus. Atoms of the same *element* have the same number of protons.

The neutron number N = the number of neutrons in the nucleus.

The nucleon number A = the number of nucleons (i.e. protons + neutrons) in the nucleus:

$$A = Z + N$$

$$^{A}_{Z}X \text{——element symbol}$$

Figure 45.4 *The nuclide symbol:*

A nuclide is a nucleus with a particular value of Z and N.

Isotopes are atoms with the *same* Z but *different* N values.

Isotopes are the same element but have slightly different masses.

Radioctivity

Properties of radiation.

	alpha particles α	**Beta particles** β	**gamma rays** γ
What it is	A helium nucleus — 2 protons — 2 neutrons	A fast-moving electron e →	Electromagnetic radiation of very high frequency and short wavelength
Charge	The charge of 2 protons: $+2e$	The charge of an electron: $-e$	No charge
Mass	Approximately the mass of 2 protons + 2 neutrons	The mass of an electron	Zero
Energy	Kinetic energy	Kinetic energy	Radiation energy in 'photons'
Deflection in a magnetic field	Small, because α has large mass: direction given by Fleming's left-hand rule	Large, because β has small mass: direction opposite to α-particles	No deflection because γ has no charge
Absorption	Maximum range: a few layers of human skin or a sheet of paper	Maximum range: a few mm of aluminium	No maximum range. 1 cm of lead absorbs about $\frac{1}{2}$ of γ-rays. Another cm halves the γ-rays again
Range in air	A few cm. They give up kinetic energy quickly forming many ions	Fast β-particles: a few metres. They give up kinetic energy gradually	Get weaker as they spread out from the source. At twice the distance they are four times weaker

Background radiation

- 37% radon and thoron gas (in houses)
- 19% γ-rays (from rocks and soil on Earth)
- 17% from radioactive elements in your body
- 14% cosmic rays (from space)
- $11\frac{1}{2}$% medical, mainly from X-rays
- $1\frac{1}{2}$% fallout, nuclear waste and industry
- (87% (the first four together) is **natural** radiation)

The cloud chamber

Radiation causes **ionisation** of a gas. This means the radiation leaves a trail of positive gas **ions** along its track.

Molecules with any electrons removed are called ions.

Water vapour molecules in the saturated air in a cloud chamber are attracted to these positive ions. They condense on the ions. Small water droplets form. They can be seen as a track when bright light is reflected from them.

Gamma radiation does not leave an actual track. Wispy tracks appear along the path of an intense γ-ray. These are caused by electrons released from atoms which cause ionisation along the path.

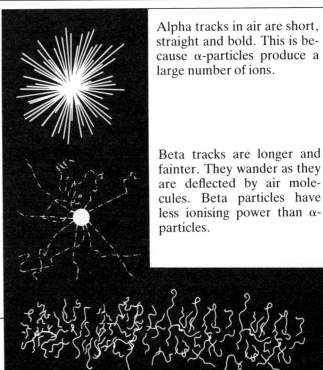

Alpha tracks in air are short, straight and bold. This is because α-particles produce a large number of ions.

Beta tracks are longer and fainter. They wander as they are deflected by air molecules. Beta particles have less ionising power than α-particles.

Figure 46.1 *Cloud chamber tracks.*

What is a half-life?

The **decay** of unstable or 'radioactive' atoms is a random event in two senses. We cannot tell:

● *which* particular atoms are going to decay;
● *when* they are going to decay.

In a large sample the random decay of individual atoms is averaged out. This gives a constant **half-life** for the particular isotope or nuclide.

> The half-life of a sample of radioactive substance is the time taken for half of the unstable atoms to decay.

How do we measure the half-life of a radioactive source?

● Use a Geiger–Müller tube and counter. This will find the count rate C from the source.
● Keep the source at a constant distance from the Geiger–Müller tube.

Background radiation produces a background count C_o. This is produced even when the source is not present.

● Subtract C_o from each reading:

corrected count rate for source only $= C - C_o$

● Plot a graph of the corrected count rate against time.

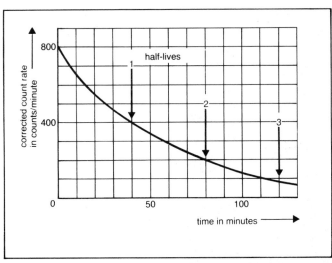

Figure 46.2 *A decay graph for a radioactive substance.*

Example: The half-life of thorium 234 is 24 days. A 32 g sample is taken. How much will remain after 96 days?

$$96 \text{ days} = 4 \times 24 \text{ days} = 4 \text{ half-lives}$$
$$\therefore \text{ mass remaining} = \tfrac{1}{2} \times \tfrac{1}{2} \times \tfrac{1}{2} \times \tfrac{1}{2} \times 32 \text{ g} = 2 \text{ g}$$

How does a nucleus change when it decays?

When a radioactive atom decays its *nucleus* throws out an α-particle or β-particle with some kinetic energy and forms a new atom of a different element. Sometimes extra energy is released as a γ-ray. These events can be described by a **nuclear equation**. In this a **parent nuclide** X changes into a **daughter nuclide** Y.

Alpha decay

$$\underset{\text{parent nuclide}}{^{A}_{Z}X} \rightarrow \underset{\text{daughter nuclide}}{^{A-4}_{Z-2}Y} + \underset{\text{alpha particle}}{^{4}_{2}\alpha}$$

An example is:

$$\underset{\text{uranium}}{^{238}_{92}U} \rightarrow \underset{\text{thorium}}{^{234}_{90}Th} + \underset{\text{alpha}}{^{4}_{2}\alpha} + \underset{\text{gamma ray}}{\gamma}$$

● A nuclide may decay by emitting an alpha particle. Then its proton number Z decreases by 2 and its nucleon number A decreases by 4.

Beta decay

$$\underset{\text{parent nuclide}}{^{A}_{Z}X} \rightarrow \underset{\text{daughter nuclide}}{^{A}_{Z+1}Y} + \underset{\text{beta particle}}{^{0}_{-1}\beta}$$

An example is:

$$\underset{\text{strontium}}{^{90}_{38}Sr} \rightarrow \underset{\text{yttrium}}{^{90}_{39}Y} + \underset{\text{beta}}{^{0}_{-1}\beta}$$

● The only change β-emission causes is to increase the proton number Z by 1.
● A neutron in the nucleus splits into a proton and an electron. The electron is emitted from the nucleus as a β-particle.

Applications

● **Tracers.** A radioactive substance can be added to a fluid in a pipeline. It is used to measure how fast the fluid flows along the pipe. It will also find leaks in the pipe. Detectors follow the radiation.
● **Flaw detection.** A large cobalt-60 source can be placed inside a steel pipe. The γ-radiation will expose a photographic plate on the outside. The radiation will show up flaws such as cracks or bubbles in welds. No power supply is needed.
● **Thickness control and measurement.** The intensity of the radiation penetrating a sheet of plastic, paper or metal can be measured. This indicates its thickness and can be used to control its thickness accurately during manufacture.

47
Nuclear energy

What is nuclear fission?

The nucleus of the uranium nuclide $^{235}_{92}U$ has a very rare property. It splits into two roughly equal halves when it captures an extra neutron.

Splitting of the nucleus is this way is known as nuclear fission.

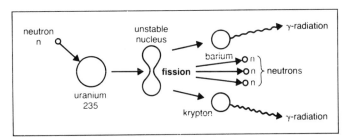

Figure 47.1 *Nuclear fission:*

The products of the fission of a uranium-235 nucleus are:

● two atoms of new elements – the fission products which are radioactive 'waste';
● two or three fast-moving free neutrons – they carry lots of kinetic energy from the fission;
● about 3.2×10^{-11} J of energy appearing as:
  (i) kinetic energy of the free neutrons;
  (ii) γ-radiation.

$$^{1}_{0}n + ^{235}_{92}U \rightarrow ^{143}_{56}Ba + ^{90}_{36}Kr + ^{1}_{0}n + ^{1}_{0}n + ^{1}_{0}n + 3.2 \times 10^{-11}J$$
of energy

a neutron collides with a uranium–235 nucleus — barium krypton (fission fragments) — neutrons

Where has this energy come from?

The total mass of the fragments and neutrons on the right-hand side of the equation is less than the total mass before fission occurred. The lost mass m appears as energy E. The energy can be calculated using Einstein's famous formula:

$$E = mc^2$$

If m is in kg and c, the speed of light, is 3×10^8 m/s, then E will be in joules.

Chain-reactions

Nuclear fission is caused by the capturing of an extra neutron. This makes a nucleus become unstable. A nuclear reactor needs to produce a continual supply of energy. So uranium-235 nuclei must always be capturing neutrons and undergoing fission. A **chain-reaction** is needed.

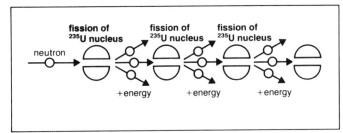

Figure 47.2 *A chain-reaction:*

In a chain-reaction each fission releases two or three extra free neutrons.

● If more than one neutron from each fission is captured by a U-235 nucleus and each results in another fission, then the chain-reaction gets out of control. This causes a nuclear explosion.
● If less than one neutron from each fission is captured by a U-235 nucleus, the chain-reaction slows down and stops.
● If, on average, one of these neutrons is captured by another U-235 nucleus, the chain-reaction will continue under control at a constant rate. This is called a **critical reaction**. In a nuclear reactor this control is maintained by the *moderator* and *control rods*.

	The moderator	**A control rod**
Action	Slows down the neutrons	Absorbs neutrons
Reason	Allows more neutrons to be absorbed by ^{235}U nuclei	Reduces the number of neutrons available for capture by ^{235}U
Effect on chain-reaction	Speeds up the chain-reaction	Slows down the chain-reaction
Material used	Graphite or pressurised water	Boron-steel or cadmium
How used	Present in the reactor core between all the fuel rods	Can be lowered into the reactor core to slow down the chain-reaction

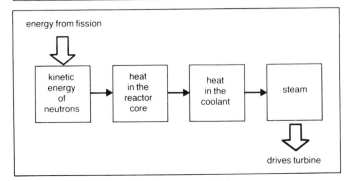

Figure 47.3 *Getting the energy out of the reactor:*

Comparing fossil fuels and nuclear power.

Fossil fuels: coal, oil and natural gas	Nuclear power: thermal and fast (neutron) reactors
Getting the fuel ● Mining and drilling are dangerous; many lives are lost ● Coal miners suffer lung diseases caused by breathing dust	● Mining uranium ore is dangerous ● Miners breath radioactive dust and radon gas. These cause diseases like lung cancer
Handling the fuel ● Coal dust blows from stock piles ● Oil spillage at sea and in harbours causes pollution. This kills wild life, birds and fish ● Oil storage tanks and coal stock piles are unsightly ● Heavy oil tankers and coal trucks are noisy. They damage roads	● Special safety precautions are needed for transporting fuel rods ● Fuel needs reprocessing after use in a reactor core ● Some risk of radioactive material leaking into the environment ● Some risk of theft for political or terrorist purposes
Using the fuel ● Produces airborne dust, smoke and gases including: sulphur oxides (causing acid rain) and radioactive thorium and potassium in the dust ● Produces a large bulk of waste ash ($20\,m^3$ of ash per person lifetime) ● Power station is safe. But it produces waste heat in the environment ● Fossil fuels have other valuable uses as chemical raw materials (e.g. for making plastics)	● Great effort must be made to limit the escape of radioactive gases and particles into the air ● Undesirable discharge of radioactive liquids into the sea from fuel reprocessing plant ● Waste products need safe storage, some for many thousands of years ● Limited risk of major disaster through fire or explosion of reactor (radioactive material escaped into the atmosphere at Chernobyl in 1986) ● Produces waste heat in the environment

The kinetic energy of the neutrons is turned into heat in the reactor core. This happens when:

● the moderator slows them down;
● the control rods absorb them.

The coolant (gas or water) is circulated around the core. As it travels it absorbs heat from the core. Heat is carried to the pipes of a steam generator. Steam at a high temperature and pressure is used to drive turbines. These generate electricity.

What is a thermal reactor?

Both the Advanced Gas-cooled Reactor (AGR) and the Pressurised Water Reactor (PWR) use a moderator. This slows down the neutrons to what are called **thermal neutron** speeds.

AGR and PWR reactors are generally known as thermal reactors. That is, their neutrons are slowed down.

What is a fast reactor?

Fast reactors use *fast* neutrons for producing energy in their cores. They have no moderator.

Fast reactors are also called **fast breeder reactors**. This is because they can be used to *breed* a new nuclear fuel called plutonium. The otherwise wasted isotope of uranium, $^{238}_{92}U$, captures a fast neutron. It becomes plutonium-239, $^{239}_{93}Pu$. This isotope of plutonium undergoes fission like uranium-235. It can be used as a fuel in fast breeder reactors. A breeder doesn't absorb spare neutrons in control rods. Instead it absorbs them in a 'blanket' of uranium-238 wrapped around its core. Such a reactor could breed its own fuel.

What is nuclear fusion?

Fusion is the 'melting' together of two atomic nuclei

Fusion is the process which powers the Sun. On Earth fusion has been achieved only in the uncontrolled explosion of the hydrogen bomb. The Joint European Torus (JET) is a research project near Oxford, where physicists and engineers are studying a possible method of controlling nuclear fusion. They hope to produce a new energy source for the future. Nuclear fusion is attractive as an energy source. This is because:

● the fuels it needs, deuterium and lithium, are in plentiful supply in the sea:
● there are no nuclear waste materials.

48
Physics and health

Safety

The effects of ionising radiation on people

When our bodies absorb energy from ionising radiation (α, β, γ or X-radiation), ions are produced which can change or destroy living cells. The radiation **dose equivalent** received by a person is measured in **sieverts**. This dose is a measure of the energy absorbed by each kilogram of the body. (This is after allowance has been made for the type of radiation).

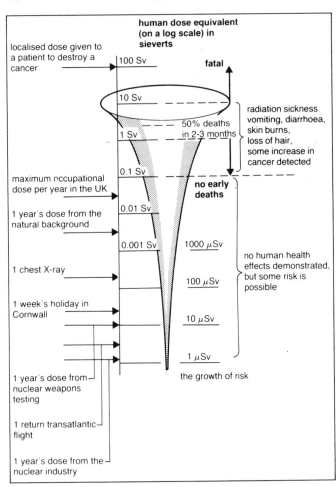

Figure 48.1 *Doses of radiation and known effects on humans:*

Our bodies are highly sensitive to all ionising radiations. They cannot repair some of the damage inflicted. Damaged cells remain in the body. They can cause cancers many years later. We assume that small doses of radiation cause very little damage. On this basis we decide what is an acceptable risk for a patient having an X-ray or for a person who uses radioactive materials at work.

How can people be protected from radiation?

The dose a person receives from X-rays or γ-rays can be limited by:
● using shielding (often made of lead);
● keeping as far away from the source as possible;
● keeping exposure times as short as possible (this is important when taking X-ray photos).
Unborn babies and young children should not be exposed to any ionising radiation.

Alpha and beta radiation are most dangerous when they come from a source which has been eaten or breathed into the lungs. This risk is limited by:
● not eating or drinking where radioactive materials are used;
● wearing disposable gloves and protective clothing. This includes masks with air filters.

Useful devices from physics

Glasses

People wear glasses to correct their faulty eyesight. The optician uses knowledge of the physics of lenses and eyes to prescribe the right lenses. Spectacles can be fitted with an ultrasound transmitter. These can be used by a blind person to estimate the distance of something in front. The delay of the echo gives the distance.

Hearing aids

Hearing aids are made for people with some forms of deafness. They depend upon electronic amplification.

The electrocardiograph or e.c.g. machine

This electronic machine tells doctors about the working of the human heart. It uses electrodes on your skin. These detect the small electrical voltage which was generated inside your heart to make its muscles contract. The shape and frequency of the electrical pulse, displayed on an oscilloscope screen, shows how well your heart is working.

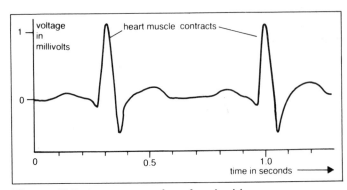

Figure 48.2 *An e.c.g. waveform for a healthy person.*

How can we see inside a human body?

Method	Radiation used		Advantages and disadvantages
X-ray imaging	X-radiation from X-ray tube is sent through the body. It casts a shadow on a photographic plate 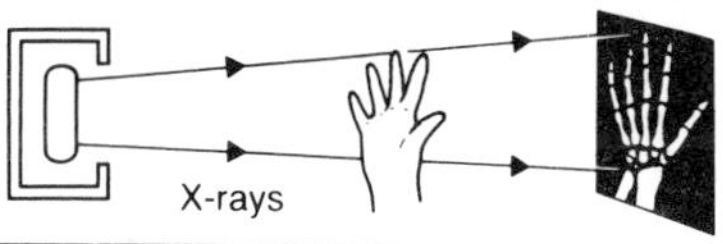		Dangerous ionising radiation, but gives sharp pictures of bone and tissue; much used
Gamma camera	Isotopes taken into the body emit γ-rays. A special γ camera detects them 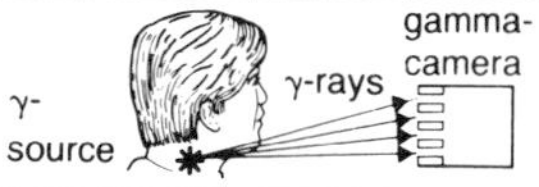		Dangerous ionising radiation. But it gives images of particular organs inside the body
Ultrasound	High-frequency sound waves reflected from things inside the body. Echo gives position, shape and texture 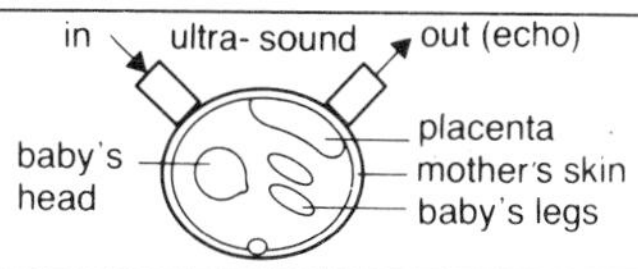		Thought to be safe, used to image unborn babies and the beating heart, but images not very sharp
Thermography	Infra-red radiation emitted from the body gives heat photographs of the body in false colours. The image is whiter where the body is hotter and emits most IR. 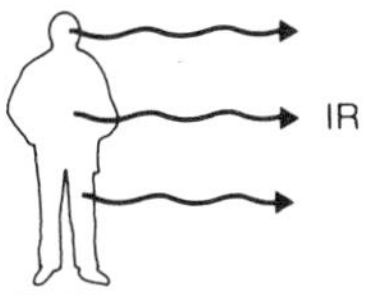		Completely safe, used to find hot or cold spots on the body; helps to find blood circulation problems and cancers
Nuclear magnetic resonance (NMR)	Coils produce a strong magnetic field. High-frequency radio waves detect protons in atomic nuclei. They show the body's structure 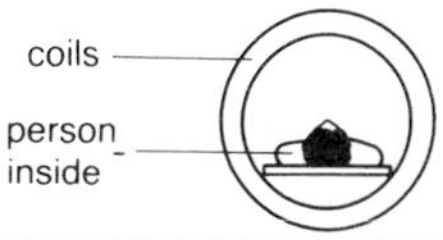		New technique, thought to be safe. It gives sharp pictures in section through the body. Equipment costs over £1 million
Optical fibre endoscopy	Light is sent down an optical fibre to light up a space inside the body. It is reflected back up a bundle of thousands of fine glass fibres to form an image 		Doctors can see inside people (e.g. lungs and stomach) and look around before operating. It is very safe and much used

Figure 48.3

Pacemakers

Some people's hearts do not beat regularly or reliably. An electronic device called a pacemaker can be fitted to them. The device generates an electrical pulse. This is fed to the heart muscles. It makes them contract and pump the blood. The electronics can control the pulse rate to suit how hard a person is working.

Treatment

Radiation treatment

Large doses of gamma radiation are used to kill cancers. The source often used is the isotope of cobalt, $^{60}_{27}$Co. A cube about the size of a sugar lump is used. It can give a dose equivalent to 100 sieverts in a short time. This dose is concentrated on the site of the cancer and destroys cancer cells. But if spread over the whole body it would kill a person. This dangerous source must be heavily shielded and operated by remote control.

Laser treatment

● A laser can be used as a surgery tool. It can be used as a knife. Then it has the advantage that the light beam is absolutely sterile. Also heat is generated which stops the bleeding where it has cut. The beam can also vaporise unwanted cancerous tissue.

● A laser is used to 'spot weld' a detached retina. (This is the light-sensitive inside of the back of the eye.) Small scars are produced when the laser beam 'welds'. But this injury allows the healing process to join the retina to the tissue behind it.

● A laser can be used to treat a port-wine birthmark. Blue/green light is produced from an argon laser. This is absorbed by the red coloured blood vessels in the birthmark. When the injury caused by the laser light heals, new tissue grows under the skin which is lighter in colour.

Review questions: Chapters 44, 45, 46, 47 and 48

C44 **1** (a) What is meant by the name *national grid*?
(b) Name two undesirable effects of power losses in cables carrying currents.
(c) State two methods of reducing these power losses?
(d) How is the power loss calculated for a cable of resistance R, current I?

2 A generator produces power of 100 kW. It is connected to a factory by a cable of 5 Ω. The output voltage of the generator is 5000 V. What is:
(a) the current flowing along the cables;
(b) the power dissipated in the cables;
(c) the voltage drop along the cables?

3 A step-down transformer at a power station has 2400 turns on its primary coil and 600 turns on its secondary coil. The input voltage is 132 kV. What is the output voltage?

C45 **4** (a) What was the name of the experiments which led to the Rutherford–Bohr model of the atom?
(b) What does this model state about the:
   (i) size of the nucleus;
   (ii) charge of the nucleus;
   (iii) electrons in the atom?

5 (a) What are the three particles in the atom?
Which particle(s):
(b) are found in the nucleus;
(c) has a neutral charge;
(d) has the smallest mass;
(e) has a positive charge?

6 What letters are used to represent the:
(a) atomic number;
(b) neutron number;
(c) nucleon number?

7 State the number of protons and neutrons in the following nuclei:
(a) $^{10}_{4}Be$; (b) $^{23}_{11}Na$; (c) $^{31}_{14}Si$; (d) $^{140}_{56}Ba$.

8 $^{12}_{6}C$ and $^{14}_{6}C$ are isotopes of carbon.
(a) How many protons, neutrons and electrons do these atoms have?
(b) In what way are these atoms different?

C46 **9** (a) What is: (i) an alpha particle;
      (ii) a beta particle; (iii) a gamma ray?
(b) Which of these types of radiation has:
   (i) a negative charge; (ii) zero mass;
   (iii) no charge; (iv) kinetic energy;
   (v) two protons and two neutrons?

10 Michael looks at tracks in a cloud chamber made by alpha particles.
(a) What effect does an α-particle have on the air molecules in the chamber?
(b) What has the water vapour condensed on to?
(c) Why do these tracks look white?
(d) Describe the appearance of the tracks.

11 Which type of radiation:
(a) is not deflected in a magnetic field?
(b) will be absorbed by paper?
(c) will only travel a few centimetres in air?
(d) will weakly ionise a gas?

12 A girl studies the absorption of gamma rays by lead. She finds that the gamma radiation is reduced to $\frac{1}{2}$ by 1 cm of lead.
(a) What is the most likely equipment that she uses to detect the gamma radiation?
(b) What fraction of gamma rays will she detect when she uses: (i) 2 cm; (ii) 3 cm of lead?

13 The figure shows the decay curve for a radioactive isotope of lead.

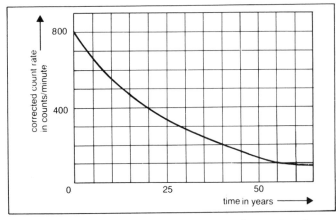

(a) What percentage of this sample remains after 40 years?
(b) What is the half-life of this isotope?
(c) Why is the 'corrected count rate' used?
(d) Name two sources of background radiation.

14 The half-life of a radioactive substance is 4 years. How long will it take for the number of unstable atoms to drop to:
(a) 50%; (b) 25%; (c) 6.25% of the original?

15 A radioactive sample of 16 g has a half-life of 6 days. How much will be left after:
(a) 6 days; (b) 12 days; (c) 24 days; (d) 30 days?

16 Write out the following nuclear equations:
(a) $^{227}_{90}\text{Th}$ changing to Ra by α-emission;
(b) $^{16}_{7}\text{N}$ changing to O by β-emission.

C47 **17** Copy and complete the following:
(a) The roughly equal splitting of nuclei is called nuclear _____.
(b) One method of starting this process is to fire very slow _____ at uranium-235.
(c) In addition to the two fragments produced there are usually two or three _____.
(d) Mass lost in the reaction appears as _____.
(e) A self-sustaining chain reaction is called a _____ reaction.

18 In a thermal nuclear reactor what is the:
(a) action of the moderator;
(b) reason for the moderator;
(c) moderator made of;
(d) name of the rods used to absorb neutrons;
(e) method of using these rods?

19 The figure shows a block diagram of the energy chain in a reactor with a steam generator. What is:

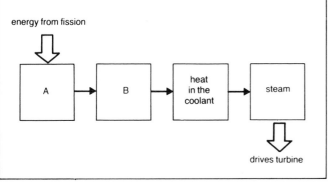

(a) the energy represented by blocks A and B;
(b) a typical material for the coolant?

20 (a) What is another name for a fast reactor?
(b) What fuel is often used in a fast reactor?
(c) Why are these reactors called 'fast'?
(d) Do fast reactors have a moderator? Why?

21 (a) What is the 'melting' together of two atomic nuclei called?
(b) Suppose this effect could be controlled in a power station. What two fuel materials could be used?

C48 **22** (a) What is produced when our bodies absorb radiation such as α- or X-radiation?
(b) What is the SI unit of the radiation dose equivalent?
(c) Which of the following help(s) to reduce the dose of radiation received:
  (i) using shielding made of lead;
  (ii) standing close to the source;
  (iii) reducing the exposure time?

23 When handling radioactive sources in a laboratory which of the following precautions should be obeyed:
(a) handling the sources with long tweezers;
(b) no eating and drinking allowed;
(c) keeping unused sources in lead-lining boxes;
(d) wearing protective clothing and gloves if the sources are strong?

24 The figure shows a typical e.c.g. waveform of a healthy person.

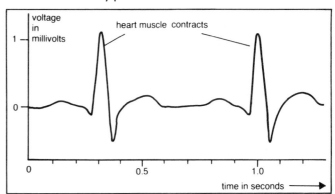

(a) What do the letters e.c.g. stand for?
(b) What does it detect?
In this example what is the:
(c) maximum voltage signal detected;
(d) number of beats/minute of this heart?

25 What type of imaging is usually used for:
(a) looking at unborn babies;
(b) producing heat photographs of the body;
(c) looking inside the lungs;
(d) studying bone and tissue?

26 (a) What is the name of an electrical device which regulates the heart muscles?
(b) What can 'spot-weld' a detached retina?

Structured and exam-type questions

C1

1 The table below gives the densities of some solids and liquids in kg/m^3.

Solids		Liquids	
Concrete	2400	Carbon tetrachloride	1630
Cork	240	Paraffin oil	800
Perspex	1190	Turpentine	870
Ice	920	Water	1000

(a) Ice floats in water. Is the density of ice smaller, the same, or greater than the density of water?

(b) From the list of solids above which would:
  (i) float in water;
  (ii) float in paraffin oil;
  (iii) sink in carbon tetrachloride;
  (iv) sink in turpentine?

2 The figure shows a grid of lines on which is drawn a closed, curved line.

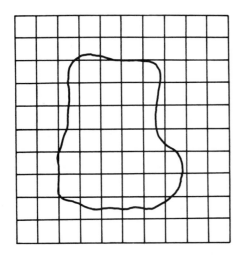

(a) Measure the length of a side of one of the small squares. Write down your answer to the nearest millimetre.

(b) Work out the area of one of the small squares, being careful to state the unit you use.

Now look at the closed, curved line.

(c) How many small squares are there inside this line? (Where there are parts of a square, try to estimate how many squares these parts add up to.)

(d) From your answers to (b) and (c), work out the area inside the closed line. Explain your working. *(NEA B)*

C2

3 A front-wheel drive car is travelling at constant velocity. The forces acting on the car are shown in the figure. *F* is the push of the air on the car.

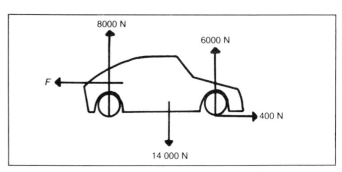

(a) Name the 400 N force to the right.
(b) What is the value of *F*, the force to the left?
(c) Taking the weight of 1 kg to be 10 N, calculate the mass of the car.
(d) The force to the right is now increased. Describe what effect this has on the speed of the car. *(LEAG A)*

4 The gravitational field strength *g* is often taken to be 10 N/kg on the Earth's surface. This is to make calculations easier. Really *g* is nearer to 9.8 N/kg on the Earth's surface. The data below give the value of *g* in N/kg at different heights *h* above the Earth's surface.

(a) Plot a graph of *g* against *h*.
  Draw a smooth curve through the points.
(b) Does the value of *g* increase, decrease or stay constant as you go further away from the Earth's surface?
(c) From your graph estimate the value of *g* at a height: (i) 1500 km; (ii) 8500 km.
(d) *g* is 9.8 N/kg on the Earth's surface. How far above the Earth's surface would you have to be for *g* to be *half* this value?

h (km)	0	100	200	700	1000	2000	3000	4000	5000	6000	7000	8000	9000
g (N/kg)	9.8	9.5	9.2	8.0	7.3	5.7	4.5	3.7	3.1	2.6	2.2	1.9	1.7

(e) (i) What is the weight of the Earth's surface of a satellite of mass 1000 kg? (Take $g = 9.8$ N/kg.)

(ii) What will be the mass of the satellite when it orbits the Earth at a height of 200 km?

(iii) What will the weight of the satellite be at this height?

C3 **5** When ball bearings are manufactured it is important that they have the correct mass. The figure shows a device for doing this.

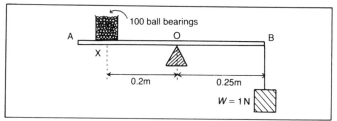

A uniform beam AB is pivoted at its centre O. It has a weight W of 1 N placed at end B, and a container of negligible weight placed at X. When the container is filled with 100 ball bearings, the beam should balance if the ball bearings have the correct mass. Assume that a mass of 1 kg has a weight of 10 N.

(a) What is the moment of the weight W about the pivot?

(b) Calculate the weight (in newtons) of the ball bearings in the container at X.

(c) What is therefore the average mass (in grams) of one ball bearing?

(d) If the container was filled with 100 ball bearings of identical size but lower density, would the end A of the beam tip up or down? Give a clear reason for your answer. (*NISEC part*)

6 A pencil and penknife can be used in a balancing trick. This is shown in the figure.

(a) Copy the diagram. Label the position of the centre of gravity of the system.

(b) Explain why the system is stable.

(c) A student removes the penknife. She tries to balance the pencil alone. Now where is the centre of gravity?

(d) Why is the pencil likely to fall over?

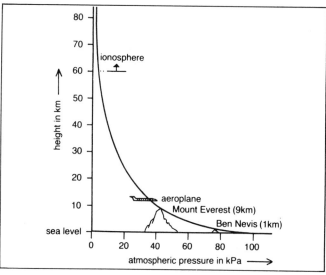

C4 **7** (a) Why does a swimmer float in water? Draw a diagram to show the forces on the swimmer.

(b) To go to the bottom of the swimming bath the swimmer often expels some air from her lungs. Why does this help her to go to the bottom of the bath?

(c) The swimmer then swims along the bottom of the bath from the shallow end to the deep end. What changes occur to the pressure on her due to the water?

8 This question is about a car braking system.

(a) What liquid is used as the brake fluid?

(b) What will happen in the system when someone tries to use the brakes if:
 (i) the system is working correctly;
 (ii) an air bubble enters the system;
 (iii) the brake fluid leaks out?

C5 **9** The figure shows a graph of atmospheric pressure against height above sea level.

(a) How does the atmospheric pressure change as you go higher above sea level?

(b) What is the pressure at: (i) sea level; (ii) the top of Mount Everest?

(c) At what height is the pressure 10 kPa?

(d) Describe the likely effect of a hole in an aeroplane flying at a height of 12 000 m.

10 A glazier uses a rubber sucker to lift a horizontal sheet of glass.

(a) Draw a diagram to show the sucker attached to the glass. Indicate the inside and outside pressures.

(b) Where is the pressure greater?

(c) Why does the seal need to be airtight?

(d) How can the glazier improve the grip of the sucker?

C6 **11** A student requires a small electric motor for a model crane. The crane load is raised on a single cord attached to a shaft driven by the motor through reduction gearing. At the start of the lift one revolution of this shaft raises the load through a vertical height of 10 mm; at the end of the lift one revolution of the shaft raises the load through a vertical height of 20 mm.

(a) Assuming that the lifting shaft rotates at a constant rate of 3 revolutions per second (rev/s), estimate the power used in lifting the load of 2 N:
  (i) at the beginning of the lift;
  (ii) at the end of the lift.
(b) The rated characteristics of three possible motors, A, B, and C are given below:

Motor	Voltage in V	Load current in A	Speed in rev/s	Efficiency
A	3	0.5	60	0.1 (10%)
B	6	0.5	60	0.1 (10%)
C	9	0.4	60	0.12 (12%)

  (i) What is the power output of each motor? (power = $V \times I$)
  (ii) The motor is to be powered from a dc source giving output voltages of 3 V, 6 V or 9 V. Given that the efficiency of the gearing system is 0.40 (40%), which motor would you recommend? Explain how you made your decision.
  (SEG)

12 An electric motor lifting heavy loads is given 3600 joules of energy/second. The pie chart shows the types of energy produced.

A: 80% mechanical   C: 10% heat due to
B: 6% heat in wires   friction at the
                                    bearings
                          D: 4% sound

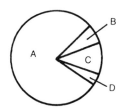

(a) What is the input power of the motor?
(b) What is the efficiency of the motor?
(c) Where is most energy lost?
(d) How many joules of sound energy are lost every minute?
(e) Suggest a method of reducing the energy lost as heat.

C7 **13** Two students are decorating a room. One uses the screwdriver to open a tin of paint. This is shown in the figure.

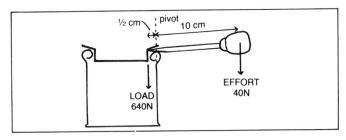

(a) What is this type of machine called?
(b) What is the value of the ratio of the effort to the load?
(c) How far does the load move if the effort moves 5 cm?
(d) By what factor is the distance moved by the effort bigger than the distance moved by the load?

The other student uses the screwdriver to loosen some screws of a fitting so they can paint behind it. This is a wheel and axle type of machine as shown below.

If the screwdriver turns once:
(e) Find the distances moved by the outside edges of the handle and blade.

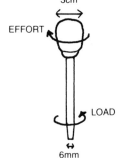

(f) What is the ratio of these distances?

14 The figure shows a wire of British Rail's electrification system, being held taut by pulleys.

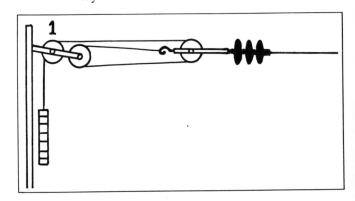

(a) By what factor is the force multiplied?
(b) What is the purpose of pulley 1?
(c) Why are pulleys used at all?

C8 **15** A walker sets off from a car park to walk along a disused railway line. The figure shows the distance–time graph of his walk.

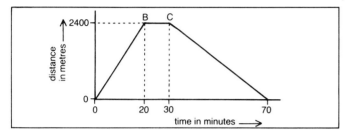

(a) He walks briskly for 20 minutes. What is his speed in m/s?
(b) How long does he rest?
(c) What is his speed during the part BC?
(d) Where does he finish his journey?
(e) How far has he walked?

16 (a) Plot a velocity–time graph of the data below of a car's motion:

t (in s)	0	10	20	30	40	50	60	70
v (in m/s)	10	15	20	20	20	20	10	0

(b) What was the initial velocity of the car?
(c) What is the maximum velocity of the car?
(d) How long does the velocity stay constant?
(e) Calculate the acceleration of the car.
(f) Calculate the deceleration of the car.
(g) How far did the car go while slowing down?

C9 **17** Slow-motion photography (see figure) shows that a jumping flea pushes against the ground for about 0.001 s, during which time its body accelerates upwards to a maximum speed of 0.8 m/s.

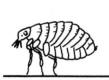

(a) Calculate the average upward acceleration of the flea's body during this period.
(b) If the flea then moves upwards with constant downward acceleration of 12 m/s², find:
  (i) how long it will take, after leaving the ground at a speed of 0.8 m/s, to reach the top of its jump;
  (ii) how high it will jump after leaving contact with the ground.
(c) Why is the acceleration of the flea after leaving the ground not equal to g?

(LEAG A)

18 A box falls from an aircraft flying at 810 m. After 6 s it has a terminal velocity of 54 m/s.
(a) How far will it fall while accelerating?
(b) How long will it take to fall the rest of the height at the terminal velocity?
(c) What is the total time of falling?
(d) Why does the box reach a terminal velocity?

On 18 September 1986 two men leapt from a hot-air balloon at over 10 000 m (35 000 feet or nearly 7 miles up) to break the world freefall record. They reached a maximum speed of 160 m/s (350 m.p.h.). Then they slowed down to a terminal speed of 54 m/s (120 m.p.h.).
(e) Explain why they fell faster at first.

C10 **19** A student carries out an experiment to find out how forces affect the motion of a trolley along a runway.
(a) Before making measurements with paper tape and a ticker timer, she tilted the runway a little. Why was this done?

She pulled the trolley along the runway using a stretched elastic thread fastened to the post P.

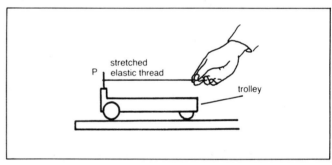

(b) What could she do to apply double force to the trolley?

After doing the experiments, the student prepared two tape charts which looked like this:

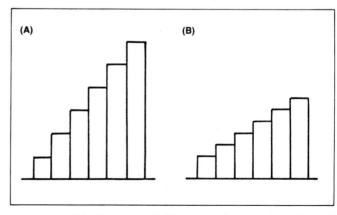

(c) How much bigger is the acceleration of the trolley in A than in B?
(d) How do the tape charts show this?

(MEG Nuff)

20 This question is about the forces experienced by a person in a lift in a department store.

  (a) The mass of the person is 60 kg. What is the person's weight?

  (b) The lift is at rest. What force is exerted by the floor of the lift on the person?

  (c) The lift now accelerates upwards at 2 m/s^2.

    (i) Explain why the person now experiences a force greater than her own weight.

    (ii) Calculate the force now exerted by the lift floor on the person.

                              (*LEAG B part*)

C11 **21** The table shows some Highway Code data of stopping distances for cars.

Initial speed (m/s)	10	15	20	25	30
Braking distance (m)	6	13.5		37.5	54

  (a) Plot a graph of the braking distance against (initial speed)^2.

  (b) Describe the graph in words.

  (c) When speed = 20 m/s, what is (speed)^2?

  (d) Use the graph to find the braking distance if the initial speed is 20 m/s.

  (e) What happens to the braking distance when (speed)^2 doubles? [Try reading off values of (speed)^2, say 100 to 200 (m/s)^2.]

  (f) How does the kinetic energy of the car:

    (i) depend on its (speed)^2;

    (ii) affect the braking distance?

22 A car is travelling at 20 m/s. It collides with a brick wall.

  (a) What is the initial velocity of the driver?

  (b) What is the final velocity of the driver?

  (c) Find the change in velocity of the driver.

  (d) The driver's mass is 75 kg. What is his change of momentum?

  (e) If the car is rigidly built and the driver is stopped by his seat-belt in 0.05 s, calculate the force exerted on him.

  (f) If the car is built so that the front end collapses gradually (i.e. a crumple zone), the driver stops in 0.1 s. What effect will this have on the force exerted on the driver?

  (g) What happens to the size of the impulse when the car has a crumple zone?

C12 **23** A hand drill has two gear wheels. The larger one, which has 56 teeth, is attached to the handle. The smaller one, which has 14 teeth, turns the drill bit.

  (a) What is the purpose of the gears on this drill?

  (b) If the larger gear wheel is turned round twice a second, how many revs per second will the drill bit do?

  (c) Suppose we wanted to make the drill bit turn faster without turning the handle any faster. How could the design of the drill be changed?

24 A manufacturer's data for the time for cars to accelerate from 0 to 60 m/s includes:
Car A = 15 s, Car B = 10 s, Car C = 12 s.
All three cars have a mass of 800 kg.

  (a) For each car, calculate:

    (i) the acceleration;

    (ii) the force with which the engine accelerates the car;

    (iii) the gain in kinetic energy of the car;

    (iv) the power of the car.

  (b) Which car has the biggest acceleration?

  (c) Which car provides the biggest force?

  (d) Which car is most powerful?

  (e) Which car is likely to have the lowest fuel consumption per metre?

C13 **25** A rocket and satellite have a mass of 2400 kg. The satellite is launched by releasing a compressed spring. If the satellite of mass 400 kg travels forwards at a speed which is 50 m/s faster than the rocket, calculate:

  (a) the mass of the rocket;

  (b) the change in momentum of the satellite;

  (c) the change in momentum of the rocket;

  (d) the change in velocity of the rocket.

26 Geostationary satellites have orbit radii of about 36 000 km. One of British Telecom's satellites has a mass of 1800 kg.

  (a) What is meant by 'geostationary'?

  (b) What is the period of this satellite?

The strength of the gravitational field at this distance is 0.3 N/kg.

  (c) What is the size of the gravitational force on this satellite?

  (d) What must the size of the centripetal force be on the satellite?

  (e) The centripetal force $F = mv^2/r$. What is the orbit speed of the satellite in km/s?

  (f) At half the orbit radius, the field strength is 1.2 N/kg. What effect does halving the orbit radius have on F and v?

C14 **27** The table below gives some information about the planets.

Planet	Distance from Sun ($\times 10^{10}$ m)	Surface gravity (N/kg)	Temperature (surface) (°C)	Density (relative to water)
Mercury	5.8	4	+350, −170	5.5
Venus	11	9	+465	5.2
Earth	15	10	+15	5.5
Mars	23	4	−23	3.9
Jupiter	78	25	−100	1.3
Saturn	143	10	−143	0.7
Uranus	287	10	−160	1.3
Neptune	450	14		1.7
Pluto	590	4	−230	0.5

Which planet(s):
(a) is 30 times further from the Sun than Earth;
(b) is almost half-way from the Sun to Pluto;
(c) has the largest value of g;
(d) has the same value of g as Earth;
(e) has the same value of g as Mercury;
(f) have densities less than water?
Explain why:
(g) the temperatures decrease with increasing distance from the Sun;
(h) Venus's carbon dioxide clouds help to make the planet hotter than Mercury.

28 The figure shows lunar eclipses.

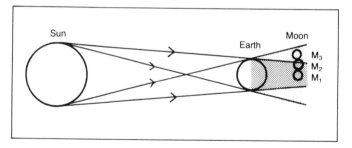

(a) What is stopping the light from the Sun reaching the Moon?
(b) What type of eclipse occurs when the Moon is in position: (i) M_1; (ii) M_2?
(c) If the Moon takes about 2 hours to pass through the umbra region, a distance of 7.2×10^6 m, what is the Moon's speed?

C15 **29** (a) Draw a ray diagram showing a drinking straw appearing bent in a glass of water.
(b) Label the angles of incidence and refraction.
(c) What happens to the speed of the light when it enters the water?

30 Six cars are parked in a car park. In daylight they look:

1	red	2	blue	3	white
4	green	5	black	6	yellow

(a) What coloured light are they reflecting?
(b) What colour will they look through a yellow filter?
(c) What colour will they look at night under a sodium (yellow) street light?

C16 **31** The figure shows a snooker table. A player hits ball P (without any spin) so that it strikes the cushion at point X. It then rebounds and hits another ball.

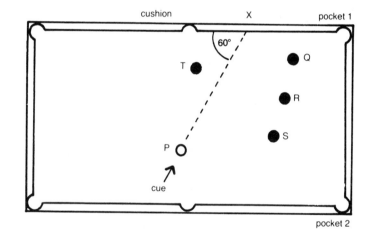

(a) Which ball will P hit?
(b) Which pocket does the ball fall into?
(c) If the angle between the direction of travel of P and the cushion is 60°, what is the angle of incidence?
(d) What is the angle of reflection?
(e) If P stops where it hits the other ball, what is true about the masses of the balls?

32 A jeweller is making a pair of silver cufflinks. She decides to produce lots of very tiny scratches on the outer part and highly polish the inner part.
(a) Describe what each part will look like when light shines on to the cufflinks.
(b) Which part gives regular reflections?
(c) What sort of reflection is produced by the other part?

C17 **33** A boy wants to show slides of his holiday. He sets up a slide projector and screen.
   (a) Draw a ray diagram to show how the enlarged image is formed on the screen.
   (b) If the picture is:
       (i) not in focus on the screen, what should he adjust on the projector?
       (ii) too large for the screen, which way should he move the projector?
       (iii) upside-down on the screen, what does he need to change?
   (c) The distance from the slide to the lens is 4 cm. From the lens to the screen it is 4 m. The image of a girl is 8 mm high on the slide. How tall will she appear on the screen?

34 (a) In a human eye which part:
       (i) controls the amount of light entering;
       (ii) alters the shape of the lens;
       (iii) refracts the light rays?
   (b) Draw ray diagrams to show an eye when it is focused on: (i) a pin; (ii) the Moon.
   (c) If the diameter of the pupil in bright sunlight is 2 mm and in twilight is 6 mm, how much easier is it for light to enter the larger pupil?
   (d) Suggest reasons why humans have two eyes.

C18 **35** (a) Copy and complete the diagram below by drawing two suitable rays to show how the converging (convex) lens forms an image of the object. Clearly mark the position of the image. (F is the principal focus.)

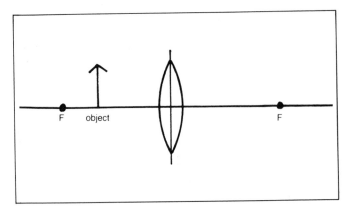

   (b) Describe the image by selecting the appropriate words. The image formed is: *real/virtual, upright/inverted* and *magnified/diminished*.
   (c) Name the optical device which uses a single converging lens to produce an image of this description.   (*NISEC part*)

36 This question compares a pinhole camera with a lens camera. Each camera is to be used to photograph a tree. Copy the figures and:

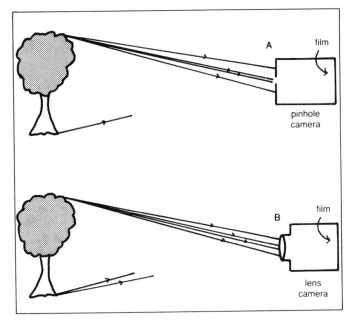

   (a) Complete the paths of any rays which go through the pinhole in A. Mark on the film the 'image' produced by the pinhole camera.
   (b) Name the type of lens used in the lens camera.
   (c) Complete the paths of any rays hitting the lens in B. Mark on the film the image produced by the lens camera.
   (d) State two ways in which the 'images' produced on the film in these cameras are different from the tree itself.
   (e) Although 'good' pictures can be obtained using a pinhole camera, most modern cameras are lens cameras. Give some advantages of the lens camera over the pinhole camera. (The figures may give you some clues.)   (*LEAG B part*)

C19 **37** *Tsunami* are tidal waves produced by earthquakes at sea. The speed of the waves depends on the depth of water, as $v^2 = gd$, where $g = 10\,\text{m/s}^2$ and d = depth.
   (a) If waves travelled at sea at 200 m/s, what is the depth of water?
   (b) If the frequency of the waves is 1 Hz, what is their wavelength at sea?
   (c) What happens to the speed of the waves when they enter shallow water near the coast?
   (d) What will then happen to the wavelength of the waves?

38 _____ waves are formed on strings when the travelling waves _____ from both ends and _____ each other. This is shown in the figure. The string is 3 metres long.

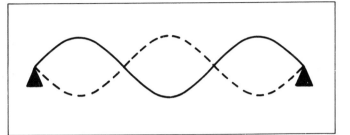

(a) What are the missing words above?
(b) How many wavelengths are shown on the string?
(c) What is the wavelength of the waves?
(d) If the speed of the waves is 40 m/s, what is their frequency?
(e) What changes could be made to the string to obtain a note of higher frequency?

C20 39 Figure P is an incomplete diagram which shows three successive straight waves A, B and C on water, as they are being reflected at a straight barrier XY. Wavefront A is just about to be reflected while B and C have already been partly reflected.

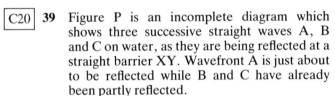

(a) Copy and complete figure P showing the positions of the reflected parts of the wavefronts B and C.
(b) Figure Q is an incomplete diagram which shows a circular wavefront originating at O just before reflection at a barrier AB. Copy and complete figure Q by drawing the wavefront as it would be just after reflection is complete. Indicate on the diagram where the reflected wavefront appears to come from.
(c) If the wavelength of an incident wave is 1.5 cm and the frequency of the source at O is 10 Hz, calculate:
  (i) the wavelength of the reflected wave;
  (ii) the speed of the waves over the water.
  (NEA A)

40 The figure shows a view from above two points P and Q. They are acting as wave sources. The waves from both sources are identical (i.e. same frequency, phase, amplitude). The sources P and Q can be changed for different experiments.

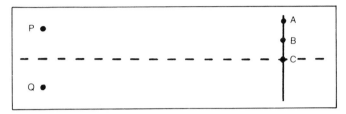

The table below shows the effects which can be obtained from the different experiments.

Waves	Sources P and Q	Effect at C	Effect at B
Water	Gaps in barrier	X	Flat water
Y	Loudspeakers	Loud sound	Quiet area
Light	Gaps in barrier	Light area	Z

(a) What words could be used to fill the spaces X, Y and Z?
(b) What is the effect at C called?
(c) What will be the effect at A?
(d) Use sketches of the two waves and their resultant to explain what happens at B.

C21 41 A waiter uses a sound pipe to communicate with the chef in the kitchen. Use a diagram to explain how the sound pipe works.

42 (a) (i) Draw a diagram to show how air carries sound from a vibrating tuning fork to the ear of a listener.
  (ii) If the fork's frequency is 250 Hz and the velocity of sound in air is 340 m/s, calculate the wavelength of the transmitted wave.
(b) The note from a tuning fork is picked up by a microphone connected to an oscilloscope. A waveform is produced on the screen.
  (i) Draw a diagram of the waveform on the screen and mark on your diagram the distance which gives the period of the waveform.
  (ii) How does the appearance of the waveform change if the tuning fork produces a louder note?
  (iii) How would the appearance of the waveform change if a tuning fork of double the frequency of the first fork were used? *(WJEC)*

C22 **43** The figure shows the tuning dials on two radios. The positions of the pointer to receive radio stations 1, 2, 3 and 4 are shown on the two dials.

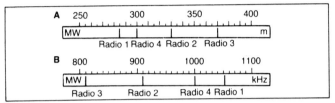

(a) Make a table giving the frequency and wavelength of the four radio stations.
(b) Use the information that you have collected to find a relationship connecting the numbers on dial A with those on dial B. Show your working.
(c) Radio 5 is found at 1100 kHz on dial B. Show the calculations that you would make to find the position of radio 5 on dial A. (*SEB*)

44 This question is about microwaves. The diagrams below show cross-sections of the oven of a microwave cooker. In Y the rotating reflector is shown after a rotation of about 180° from the position in X.

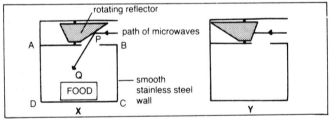

Electromagnetic waves of frequency about 2500 MHz (i.e. 2.5×10^9 Hz) enter the oven as shown. Water molecules within the food absorb the waves and the food becomes hot. Microwaves are reflected by metal but pass through glass, china, dry paper and cardboard with little change.
(a) The microwaves are reflected by the wall of the oven. Why?
(b) The microwaves of path PQ would be reflected by the walls many times. Copy the diagram of the lower compartment ABCD and draw in the path of the microwave to include the first **three** reflections.
(c) Why does this cooker use a rotating reflector?
(d) What happens to the amplitude of the microwaves as they pass through food. Why?
(e) A microwave oven uses less energy to cook food than an ordinary gas or electric oven. Why?

(f) Microwaves travel more slowly in food than in air. What effect will this have on the wavelength? (*SEG alt part*)

C23 **45** An aircraft flies just below a negatively charged thunder cloud. Movement of free electrons causes electrostatic charges to be induced in the aircraft.
(a) Copy the figure and show the positions and signs of the induced charges on the aircraft.

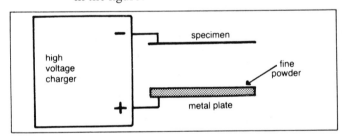

(b) Explain, in terms of the movement of electrons, the distribution of the charges you have shown.
(c) What will happen to the induced charges when the aircraft flies away from the cloud? (*LEAG A part*)

46 The police use electrostatic effects to reveal finger prints on paper cheques. This is shown in the figure.

(a) What is the charge of the powder on the plate?
(b) Why is the powder repelled from the plate?
(c) The powder remains stuck only to the tacky ridges of the finger print. Why does the powder leave the other parts of the paper and make the finger print visible?

C24 **47** A student is given a sealed box with three identical lamps. They are labelled L_1, L_2 and L_3. There are also two terminals. He connects the terminals to a 5 V power supply. Then he tries to find out how the lamps are connected inside the box. He records his actions as:

Action	L_1	L_2	L_3
Terminals to 5 V	Bright	Bright	Very bright
Unscrew L_1 only	Out	Out	Very bright
Unscrew L_2 only	Out	Out	Very bright
Unscrew L_3 only	Bright	Bright	Out

How must the lamps be connected?

48 A series circuit contains three 3 V cells and two identical lamps A and B. What is the voltage across:
(a) the 3 cells; (b) lamp A; (c) lamp B;
(d) both lamps?

C25 **49** The figure illustrates the system used for measuring the volume of fuel in a car's tank.

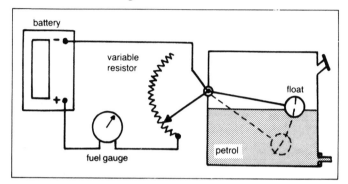

(a) Explain briefly how the system works.
(b) Describe how the scale of the fuel gauge could be marked off (calibrated) in litres.   *(SEB)*

50 For the circuit shown, what is the:
(a) resistance between A and B;
(b) resultant resistance between B and C;
(c) total resistance between A and C;

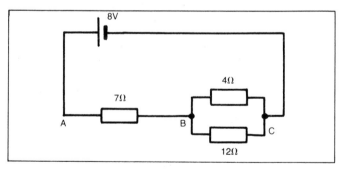

(d) current flowing from A to B;
(e) p.d. (or voltage) between A and B;
(f) p.d. between B and C;
(g) current flowing through the:
(i) $4\,\Omega$ resistor; (ii) $12\,\Omega$ resistor?

C26 **51** A student uses a bar magnet to pick up some steel needles.
(a) Why are the needles attracted to the magnet?
(b) What type of magnetic pole will be induced on the end of the needles nearest to the north pole of the magnet?
(c) Explain why the needles spread out at the end furthest from the bar magnet.

52 The figure shows some plotting compasses around a bar magnet. One of the compass needles has been drawn in. Copy the figure.

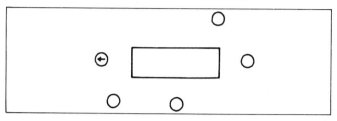

Use your knowledge of the magnetic field to:
(a) label the poles of the magnet;
(b) draw the other compass needles in the blank circles.   *(LEAG A)*

C27 **53** Two metal rods are placed in a long coil as shown. When a direct current flows through the coil, the rods move apart. When the current is switched off, the rods return to their original positions.

(a) Why do the rods move apart?
(b) From what metal are the rods likely to be made? Give a reason for your answer.

(c) If alternating current from a mains transformer is passed through the coil, what effect, if any, will this have on the rods? Explain your answer.   *(MEG)*

54 For an electric motor rated at 240 V, 720 W:
(a) what current would flow through its coils when the motor is running at full power;
(b) what fuse should be fitted in the plug?
If the total resistance of the coils is $16\,\Omega$:
(c) how much power is wasted in the coils as heat;
(d) how efficient is this electric motor;
(e) how much energy would be needed to run this motor for 1 minute?

C28 **55** A girl spends an evening at home and uses the following electrical appliances:
    a 8 kW cooker for 1 hour;
    a 3 kW immersion heater for 40 min;
    a 960 W hairdrier for 20 min.
(a) How many kWh are used for each appliance?
(b) If each kWh costs 5p, what is the cost of running each appliance?
(c) What is the total cost?

56 The figure represents a domestic electric ring main circuit. A loop of cable runs from the consumer unit round the house (or part of it) and returns to the consumer unit.

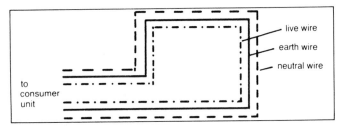

(a) Copy the figure and show how a power socket could be connected.
(b) What rating would the fuse for a typical ring main circuit have?
(c) Some household appliances such as electric cookers are connected to a radial circuit.
　(i) How does this type of circuit differ from a ring circuit?
　(ii) Explain why the diameter of the wires in a cooker is greater than that of a ring main cable.
　(iii) In a normal domestic plug one wire does not normally carry current. Which wire is it and why is it there?
(NISEC part)

C29　**57** The figure shows the structure of a bicycle dynamo.

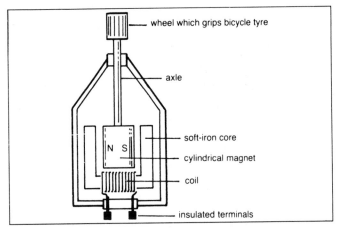

(a) As the wheels turn the axle which other part of the dynamo rotates?
(b) What happens in the soft-iron core when the wheels turn?
(c) Where will there be an induced voltage?
(d) How could a larger voltage be produced?
(e) What type of current is produced by this type of dynamo? Why?
(f) Explain why no current is induced when the bicycle wheels do not turn.

58 The figure shows part of a tape recorder. The tape has a magnetic material coated on one surface. The 'recording head' can magnetise this material.

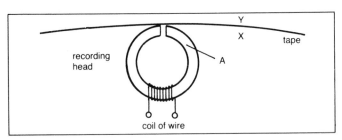

(a) Suggest a material which would be suitable for the part of the recording head labelled 'A'.
(b) Explain why you think the material you suggest would be suitable.
(c) Should the magnetic material be on side X of the tape, or side Y?
(d) Suggest how you could magnetise the tape more strongly as it passes over the recording head.
(e) Explain your answer to part (d). *(NEA B)*

C30　**59** The figure shows the screen, Y-gain and time-base controls from a typical oscilloscope displaying a waveform.

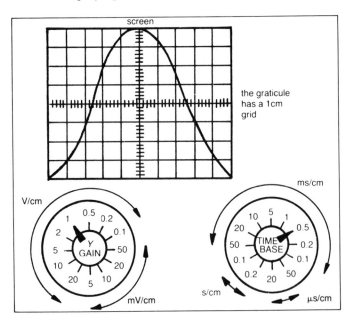

(a) What is the setting of the Y-gain control?
(b) What is the peak voltage of the waveform?
(c) What is the time-base setting?
(d) What is the period of the trace?
(e) What is the frequency of the waveform?

(f) If the time-base is altered to 1 ms/cm and the Y-gain to 2 V/cm, draw the resultant trace on a graticule like the one shown.

(NEA A)

60 An oscilloscope is used to measure an a.c. power supply. The trace on the screen is a vertical line of length 1.4 cm. The Y-sensitivity is set to 2 V/cm.
(a) Is the time-base switched on or off?
(b) What is the peak-to-peak voltage?
(c) What is the peak voltage?
(d) Calculate the r.m.s. voltage if:
peak voltage = 1.4 × r.m.s. voltage.

C31 **61** The graph shows the change of resistance of a thermistor with temperature.
(a) What is the resistance of the thermistor at 22 °C?

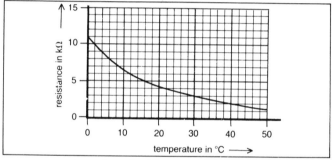

A student decides to make a thermometer and puts the thermistor in a potential divider circuit as shown.
(b) What will be the reading on the voltmeter?
(c) The voltmeter reads 4 V. What is:
(i) the resistance of the thermistor;
(ii) the temperature?

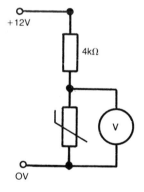

(d) Repeat (c) for a reading of 8 V.
(e) Describe how the student should mark the meter to calibrate it in °C.

62 (a) Draw a circuit diagram to show how four diodes can be used to produce full-wave rectification of an alternating current.
(b) Modify your circuit to smooth the output.
(c) Sketch graphs to show how the current varies with time:
(i) before rectification;
(ii) after rectification without smoothing;
(iii) after rectification with smoothing.

C32 **63** A transistor may be used as an amplifier using the circuit below.

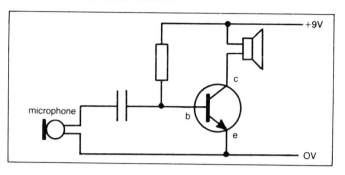

(a) What energy change occurs in:
(i) the microphone; (ii) the loudspeaker?
(b) Explain why a small varying current in the base circuit causes the loudspeaker to work.

(WJEC part)

64 The figure shows a transistor switching circuit. When switch S is opened there is a delay before the bell rings.

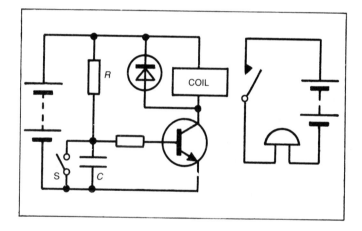

(a) What voltage is needed on the base of the transistor to turn it on?
(b) Explain why there is a delay before the bell rings.
(c) How could the components be changed so that there is a longer delay?
(d) What happens if switch S is now closed?

C33 **65** A student builds up an inverting op-amp with a voltage gain of −20. The power supply used is ±15 V.
(a) What will be the output voltage, V_{out}, if the input voltage, V_{in}, is:
(i) +0.5 V; (ii) −2 V?
(b) What is the maximum possible V_{out}?
(c) What is the minimum V_{in} to give this maximum V_{out}?
(d) If V_{in} is an a.c. signal of amplitude ±2 V, describe V_{out}.

66 The frequency response of an op-amp and an inverting op-amp circuit are shown below.

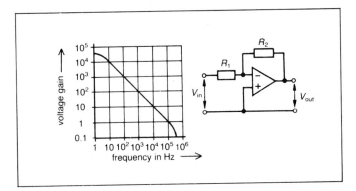

(a) $R_1 = 100\,k\Omega$ and $R_2 = 10\,k\Omega$. What is the voltage gain of the inverting amplifier?

(b) Copy the graph. Sketch on the same axes the gain–frequency graph for this inverting amplifier.

(c) What is the bandwidth of this inverting amplifier?

(d) What is the maximum voltage gain that can be obtained if the bandwidth is to be 1 kHz?

(e) Why is an amplifier with a wide bandwidth used in a hi-fi system?

C34 **67** In this circuit the relay coil is energised and can close switch S if the output of the NOT gate is high.

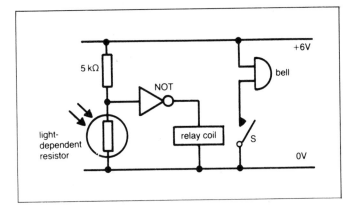

When a bright light is shone on the light-dependent resistor the input voltage to the NOT gate falls.

(a) Why does this happen?

(b) What happens in the rest of the circuit as a result of this?

(c) Suggest a practical use for this circuit. *(MEG)*

68 (a) Complete a truth table for the set of gates in the figure below.

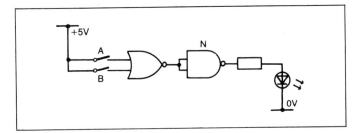

(b) Describe in words the input conditions to light the LED.

(c) Which one gate could replace this set?

This gate labelled N is replaced by two NOR gates connected as a bistable circuit.

(d) Describe how the circuit behaves now.

(e) Draw a circuit diagram of the bistable.

C35 **69** The figure shows a block diagram of a satellite radio communication system.

(a) Copy and complete the table of the parts of the system and their purpose.

(b) State the form of the signal at the points labelled W, X, Y and Z.

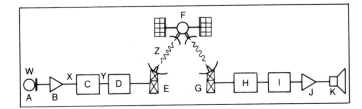

Part of system		Purpose
A		Converts sound into electrical signal
B		Increase signal strength
C	Encoder	Converts analogue signal into digital pulses
D	Transmitter	Adds digital pulses to microwave carrier
E		Aims a beam of microwaves at F
F		
G		
H	Receiver	
I		
J		
K	Loudspeaker	

70 Two uses of telephone systems are the transmission of speech and computer data.
(a) What input and output devices are used?
(b) Name three different types of links used to transmit the information.
(c) How have telephones changed our lives?

C36 **71** This question is about stretching a spring. A loaded spring is mounted vertically as shown in the figure. *h* is the height of the bottom of the load from the bench.
(a) Describe how you would use a metre rule to measure *h*. Include the precautions you would take to make your results as reliable as possible.
(b) A student measured values of *h* for several different values of load. The measurements are shown in the table:

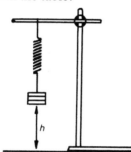

Load in N	Height *h* in mm
1	184
2	172
3	162
4	150
5	141

(i) Plot a graph of *h* (*y*-axis) against load (*x*-axis). Use a scale of 140 to 190 mm for *h* and 0 to 5 N for load.
(ii) Draw the best straight line.
(iii) Use the graph to find the load which gives a value of *h* of 180 mm.
(iv) Use your graph to find the value of *h* at a load of 1.50 N.
(v) Use the graph to help you calculate the **change in load** which gives a **change in *h*** of 1.00 mm. (*SEG*)

72 (a) Sketch an extension–force graph for a rubber band.
(b) Does the rubber band obey Hooke's law? Explain how this is shown on the graph.
(c) On the same axes add graphs for:
(i) a rubber band of half the length;
(ii) two rubber bands of the original length joined together in parallel (side by side).
(d) Explain both of these graphs.

C37 **73** The temperature of the water in a hot water cylinder is controlled by a device called a thermostat. X and Y form a bimetallic strip. Z is an electrical insulator (see figure).

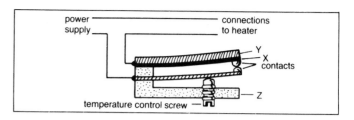

(a) Name a suitable material for Z.
(b) Explain why the bimetallic strip bends when it becomes hot.
(c) Explain what effect this has on the current through the immersion heater.
(d) If the temperature control screw is turned so that it moves to the left, does the temperature of the water increase, decrease or stay the same? (*NEA A part*)

74 40 kg of water in the hot water tank shown opposite was heated by an electrical immersion heater.

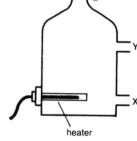

(a) To which point, X, Y or Z, is each of the following connected:
(i) the cold-water supply;
(ii) the hot taps;
(iii) the overflow pipe?
When the immersion heater was switched on, readings were taken of the average temperature of the water at different times. The following graph was obtained.

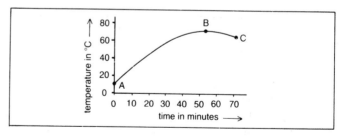

(b) The heater was switched off at B. Why is the graph from A to B not a straight line?
(c) Use the graph to calculate:
(i) the initial rate of rise of temperature of the water;
(ii) the rate at which energy was supplied to the water (1 kg of water requires 4200 J to raise its temperature by 1 °C);
(iii) the power of the heater.
(d) What would you do to reduce the heat loss from the hot water tank? Explain why this would be a good idea. (*WJEC*)

C38 **75** A refrigerator is to be used to make ice cubes. A tray is filled with 100 g of water at 18 °C. Calculate:
(a) the amount of heat to be removed to cool the water to 0 °C;
(b) the amount of latent heat to be removed to freeze this water (at 0 °C);
(c) the total amount of heat removed by the refrigerator.
(Specific heat capacity of water = 4200 J/(kg K) and latent heat of fusion of ice = 340 000 J/kg.)

76 Explain the following:
(a) When a hair drier is used to dry wet hair, at first the air from the hair drier feels cool.
(b) Breathing on a mirror makes it misty.

C39 **77** This is an experiment with air. The air in the glass tube was gradually compressed in the apparatus shown in the figure.

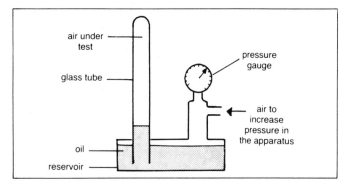

The readings taken during the experiment are shown in the table.

(a) Plot the values on a graph with volume vertical and pressure horizontal. Use a vertical scale of 0 to 100 cm³ and a horizontal scale of 100 to 300 kN/m².

Pressure in kN/m²	Volume in cm³
100	95
140	68
170	—
210	45
240	40
260	37
300	32

(b) Draw a curve to the points you have plotted.
(c) Using your graph, draw lines to the axes to find the volume of air at a pressure of 170 kN/m. What is the volume in cm³?
(d) At the start of the experiment, the oil in the tube and reservoir are not level. The gauge, which is correct, does not read zero. Why not?   (*SEG alt*)

78 Explain in simple langule how the kinetic theory accounts for the:
(a) pressure exerted by the molecules of a gas;
(b) energy required to melt a solid.   (*SEG*)

C40 **79** Figure P illustrates an instrument used to measure the time that the Sun shines during a day. The blackened glass tube contains mercury and is supported inside an evacuated glass case. Figure Q shows how the connecting wires are arranged inside tube A.

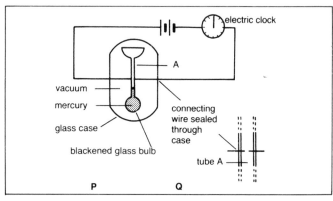

(a) How does energy from the Sun reach the mercury? Give a reason for your answer.
(b) Explain why the clock starts when the Sun shines.
(c) Why is tube A of small cross-sectional area?
(d) Explain why blackening the bulb ensures that the mercury level falls rapidly when the Sun ceases to shine.   (*MEG*)

80 The figure shows the percentage heat losses from a house.

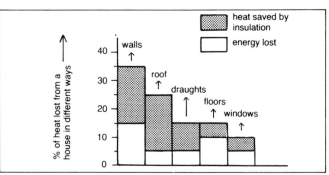

(a) Without insulation, what percentage of heat is lost through:
  (i) the walls; (ii) the windows?
(b) For each of the five methods shown in the chart, find the percentage of heat loss which could be saved by using insulation.
(c) What is the total percentage which could be saved?

(d) If the energy used by a house per day is 400 megajoules, how much energy could be saved per day if all five methods were used?

C41 **81** The mass of a girl is 45 kg.
(a) What is her weight in newtons?
She climbs a mountain and gains gravitational potential energy. This amount of energy would just be provided by one slice of bread when fully digested. She thought about this and decided that, in practice, she would probably need to eat at least five slices.
(b) Suggest one reason why this was sensible.
(MEG Nuff)

82 Energy is both vital and dangerous to life.
(a) Name two ways in which our bodies:
  (i) gain energy;
  (ii) lose energy;
  (iii) can be damaged by energy.
(b) When we run a race we need to lose heat more quickly. How does the body increase its rate of loss of heat?
(c) When we are very cold how does our body attempt to reduce loss of heat?

C42 **83** The figure shows a solar panel used for heating water.

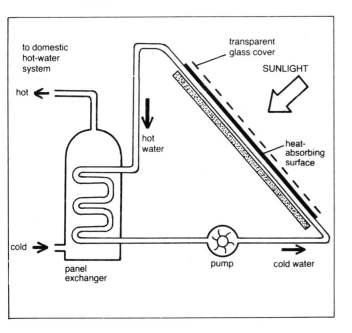

(a) Why is the surface covered with a transparent glass sheet?
(b) What colour should the heat-absorbing surface be? Why?

(c) If no pump was available, how could the water be made to circulate naturally?
(d) Why is the surface of the panel tilted at an angle to the horizontal?
(e) What is the material behind the panel?

84 Wind generators can be used to produce electrical power. The graph shows how the wind power available for each square metre of generator blade varies during the year.

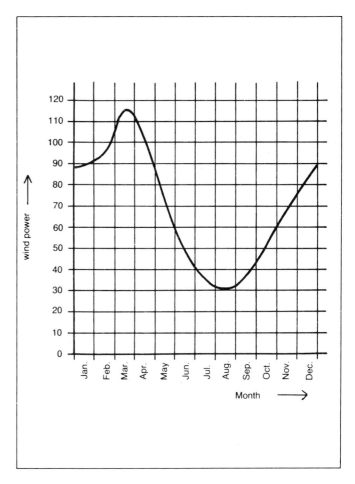

(a) In which month is the wind power greatest?
(b) Name the months of the year in which the wind power is greater than 60 units.
(c) (i) What is the power during the least windy month?
  (ii) The electrical power produced by a wind generator depends on the wind power for each square metre of the generator blades. When the wind power is 50 units for each square metre of blades, what area of blades is needed to produce 1000 units of electrical power? Show your working.
(SEB)

C43 **85** The figure shows a table decoration. The lightweight metal fan turns steadily when the candle is burning.

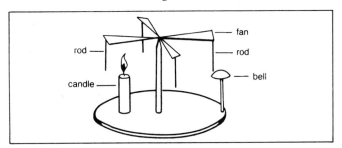

(a) Explain why the fan turns. A metal rod hangs loosely from the end of each blade of the fan and strikes a bell as the fan turns.

(b) Starting with the candle, state the energy transfers which occur.   (*MEG Nuff*)

86 In a motor car only a small percentage of the energy stored in the petrol is converted into kinetic energy.

(a) What other forms of energy are produced?

(b) By what processes are they produced?

C44 **87** The figure shows a power station which generates 100 MW of power. The voltage is stepped up to 400 kV and then the power is transmitted by the national grid over a large distance. The voltage is then stepped down before the power is used by industry and homes in a town.

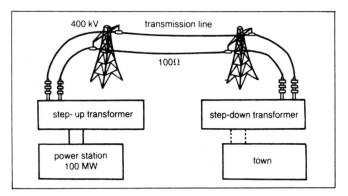

(a) Given that 100 MW is fed into the transmission line at 400 kV, calculate the current flowing in the transmission line.

(b) If the total resistance of the transmission line is 100 Ω, calculate the potential drop along the line due to the current.

(c) Calculate the power 'lost' along the transmission line.

(d) Calculate what fraction of the power is 'lost'.

(e) What happens to this 'lost' power?

(f) Explain why less power is 'lost' when a given amount of power is transmitted at high voltage and low current rather than high current and low voltage.

(g) Why is there a saving in the cost of cables when the current is low?

(h) The electricity board normally transmits power over large distances using overhead power lines but the general public would often prefer power lines to be underground.

(i) Give one advantage of having overhead power lines.

(ii) Give one advantage to the general public of having the power lines underground.   (*LEAG B*)

C45 **88** When alpha particles are fired at a thin gold foil, most pass straight through and only a very few are deflected back towards the source.

(a) What does this mean about the size of the nucleus of the gold atom?

(b) What particles are in the nucleus?

(c) Gold can be expressed as $^{197}_{79}$Au. How many particles are in the nucleus?

(d) What does this mean about the size of the particles in the nucleus?

(e) What other particles are in the atom?

(f) How many of these other particles are in the gold atom?

C46 **89** This question is about radioactivity. The graph shows the variation in activity of a sample of an isotope of radon.

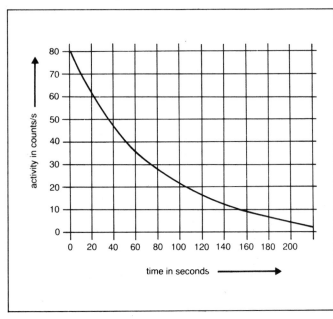

(a) Explain what is meant by the terms:

(i) activity; (ii) isotope.

(b) Using the graph, obtain a value for the half-life of the sample. Explain clearly how you obtained your value.

The figure shows the tracks produced in a diffusion cloud chamber by a radioactive source.

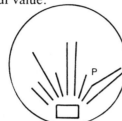

(c) (i) What type of radiation could produce tracks like this? Give two reasons.
   (ii) Describe what might have happened at P and explain your answer.
(d) An element X decays by giving off an alpha particle. Copy and complete the equation below, showing what took place.
$$^{238}_{92}X = ^{---}_{--}Y + ^{---}_{--}$$
(SEG alt)

90 A company which manufactures soap wishes to check the level of soap powder packed in boxes. It proposes to place a radioactive source on one side and a suitable detector on the other side of the conveyor belt which carries the boxes, as shown in the figure.

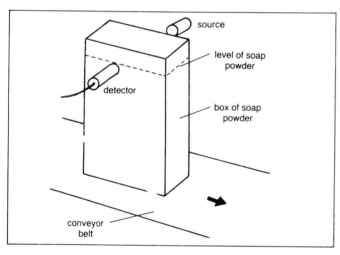

A catalogue of sources lists the following information about available sources:

Source	Radiation emitted	Half-life
A	Alpha	5000 years
B	Alpha and gamma	4000 million years
C	Beta	5 days
D	Beta	30 years
E	Gamma	20 minutes

You are asked to advise the company on which radioactive source to use.
(a) Which type of emitter would you choose? Explain your answer.
(b) In what way does the half-life of a source affect your choice?
(c) Which source in the list above would you recommend the company to use?
(d) Explain how the system will allow the correct level of soap powder to be checked.
(e) What precautions should the company take to protect people from the radiation from the source? *(NEA B part)*

C47 91 People concerned about the safety of nuclear power stations might ask the questions below.
   Suggest some answers to each question. Explain your answers:
(a) Can people work at a nuclear power station with no threat to their health?
(b) Are there any potential hazards to people who live near nuclear power stations?
(c) Is any environmental pollution caused by nuclear power stations?
(d) What happens to the waste products from a nuclear power station?

92 Describe with reasons what would happen in a nuclear reactor if:
(a) the control rods were all fully inserted into the reactor core;
(b) the control rods were removed from the reactor core;
(c) the core had no moderator;
(d) the coolant leaked away?

C48 93 Radiation can be both dangerous and helpful when it meets a human body.
(a) What types of radiation are dangerous to people? Why?
(b) What are three ways of protecting people who need to use radiation in their work?
(c) Describe two ways in which radiation can be used to benefit a person under medical care. What type of radiation is used in each case?

94 (a) List two medical applications of physics in each of the following:
   (i) medical diagnosis;
   (ii) medical treatment;
   (iii) devices which help people with handicaps.
(b) Describe one example in more detail.

Answers to review questions

Chapters 1, 2 and 3

1 (a) the amount of space it occupies.
   (b) the amount of matter in it.
   (c) the mass of its atoms and how closely packed they are; or mass and volume.
   (d) the mass of a unit volume of it.
   (e) clocks

2 (a) m (b) s (c) m^3 (d) kg (e) m
   (f) kg/m^3 (g) m^2

3 (a) tape measure
   (b) micrometer
   (c) ruler
   (d) pair of scales or beam balance
   (e) micrometer

4 (a) m, 10^{-3} or 0.001
   (b) μ, 10^{-6} or 0.000001
   (c) k, 10^3 or 1000
   (d) M, 10^6 or 1000000
   (e) c, 10^{-2} or 0.01
   (f) n, 10^{-9} or 0.000000001
   (g) G, 10^9 or 1000000000
   (h) p, 10^{-12} or 0.000000000001

5 (a) 10 (b) 40 (c) 500 (d) 1000
   (e) 1200

6 (a) 4000000 W (b) 0.003 m
   (c) 0.000001 A (d) 0.06 m
   (e) 7300 Hz

7 (a) 5 kJ (b) 2 cm or 20 mm
   (c) 4 mg (d) 1 μs

8 (a) 2×10^6 mm^2 or 2000000 mm^2
   (b) 4×10^4 cm^2 or 40000 cm^2
   (c) 3×10^{-6} m^2 or 0.000003 m^2
   (d) 1×10^{-9} m^3 or 0.000000001 m^3

   (e) 8×10^{-6} m^3 or 0.000008 m^3
   (f) 5×10^9 mm^3 or 5000000000 mm^3

9 1.2 g/cm^3 or 1200 kg/m^3

10 5.25 g or 5.25×10^{-3} kg

11 (a) Up = upthrust from helicopter blades
        Down = weight, W
    (b) Right = force from engine and friction between tyre and road
        Left = air resistance, R
    (c) Up = force of table on bowl
        Down = weight of bowl, W
    (d) Up = normal force
        Down = weight, W
        Right = friction force towards wall, F
        Left = contact force, C

12 (a) size (magnitude) and direction
    (b) size (magnitude) but no direction
    (c) newton, N
    (d) arrow
    (e) g, N/kg

13 b, c and g

14 500 N

15 W = weight = force downwards in N, m = mass in kg, g = gravitational field strength in N/kg
    (a) $m = 20$ kg on Earth and on Pluto
    (b) $W = 200$ N on Earth, $W = 80$ N on Pluto

16 (a) 2 kg (b) 20 N (c) 20 N

17 180 N, 150 N

18 (a) 78 N (b) 40° to horizontal

19 (a) turning (b) fulcrum.
    (c) perpendicular distance from the pivot to the line of action of the force.
    (d) 90° or a right angle.

20 (a) the hinge of the door
    (b) 45 N m (change 75 cm into m)

21 (a) CG = approx centre of basket
    (b) See figure 3.3 bottom right, page 6
    (c) CG = centre of board half way along
    (d) CG = centre of sugar lump
    (e) CG = in space, on the central axis, towards the fatter end

22 (a) 800 N m (b) 800 N m (c) 320 N

23 (a) returns to its upright position.
    (b) more (c) less (d) below or under

24 (a) pear sits on opposite end to stalk
    (b) pear rests on stalk end
    (c) pear lies on its side

25 (a) Hyacinths in bowl have a low centre of mass, a wide base and are stable. Daffodils in tall vase have a higher centre of mass, a smaller base, are unstable and likely to fall over if knocked.
    (b) The tendency of a yacht to capsize during fast sailing action is counterbalanced by the crew leaning outboard on the opposite side. This extends the stability range of the yacht.

Chapters 4, 5, 6 and 7

1 (a) normal
   (b) pascal, Pa or N/m^2
   (c) increases
   (d) area (or shape)

2 2000 N/m^2 or 2000 Pa

3 (a) With castor cups, force of chair acts over larger area. So it exerts a smaller pressure
   (b) Area of toe is less than area of

foot. Force of body = weight, which stays the same. So pressure on floor is greater when standing on toe.

4 61 800 Pa or 61.8 kPa

5   800 Pa (change 10 cm to m)

6   13.5 kN

7   (a) fluid displaced.
    (b) weight, upthrust or buoyancy force.
    (c) liquid

8   (a) 42 kPa (change 30 cm to m)
    (b) 142 kPa

9   (a) The height of mercury in the tube will reduce (atmosphere supports less liquid)
    (b) 760 mm = 76 cm = 0.76 m
    (c) 103 600 Pa

10 (a) push plunger in, put barrel into water, pull plunger out of barrel
    (b) air pressure > pressure in barrel

11 (a) air pressure, usually calibrated in height above sea level
    (b) aneroid barometer

12 (a) direction
    (b) less
    (c) work
    (d) potential

13 (a) 120 J (b) 45 J

14 (a) 10 kW (b) 8 kW (c) 4 kN
    (d) 240 kJ (e) 60 s

15 (a) 4200 J (b) 840 W

16 72 kJ (change 20 minutes to seconds)

17 (a) 100 kJ (b) 100 kJ

18 (a) greater (b) less (c) number

19 (a) distance (b) force (c) force
    (d) distance (e) force (f) distance

20 (a) 4
    (b) effect of friction and weight of pulley block
    (c) at large loads the weight of the pulley and the frictional forces are small compared to the load, so the efficiency rises.

21 (a) 4 (b) force

Chapters 8 and 9

1   (a) distance
    (b) instantaneous
    (c) 50 dots
    (d) gradient or slope

2   a, b, c, e, f and h

3   (a) (i) 54 km/h (ii) 15 m/s
    (b) (i) 18 km/h (ii) 5 m/s
    (c) (i) 72 km/h (ii) 20 m/s

4   (a) 3600 m
    (b) 108 000 m or 108 km

5   (a) 10 (b) 0.2 s (c) 60 mm or 6 cm
    (d) (i) 300 mm/s (ii) 0.3 m/s

6   (a) $v_1 = 60$ mm/s, $v_2 = 45$ mm/s
    (b) friction

7   (a) ticker-tape
    (b) ticker-timer

    (c) tick
    (d) ten-tick
    (e) decreasing speed
    (f) all strips equal length

8   (a) 10 cm (b) 1 s (c) 10 cm/s

9   (a) (i) straight line through origin, gentle slope
        (ii) straight line through origin, steep slope
    (b) stopped moving, so constant distance from start

10 (a) low constant acceleration from rest
    (b) high constant acceleration from rest
    (c) zero speed
    (d) low constant speed
    (e) high constant speed
    (f) constant deceleration from an initial speed

11 2 m/s ($\Delta s/\Delta t = (6-2)/(5-3) = 4/2 = 2$)

12 (a) 2 m/s^2 (b) 5 s (c) 1 m/s^2
    (d) 400 m

13 b, d and e

14 (a) 1 m/s^2 (b) 2 m/s^2

15 (a) 10 m/s (b) 24 m (c) 6 m/s, 30 m

16 (a) 10 m/s^2 (b) 20 m/s (c) 20 m

17 (a) 4 s
    (b) zero
    (c) 40 m/s upwards
    (d) 80 m

18 (a) more than two-ticks gives very long tape
    (b) 0.04 s
    (c) 0.4 s
    (d) $v = 380$ cm/s so $g = 950$ cm/s^2
    (e) 9.5 m/s^2, friction (drag of tape going through timer)

Chapters 10, 11 and 12

1   (a) direct
    (b) inverse
    (c) unbalanced, 1 m/s^2
    (d) unbalanced, unbalanced

2   (a) 4 N
    (b) increase (resultant force)

3   (a) 600 N (b) 360 N (c) 240 N
    (d) terminal

4   (a) opposite
    (b) reaction, 100 N
    (c) tree falls over

5   (a) 2500 J

    (b) 2×10^{-3} J (change 4 g to kg)
    (c) 5×10^{-9} J (change 1 mm/s to m/s)

6   (a) $2^2 = 4$ times
    (b) $3^2 = 9$ times

7   (a) 700 N (b) 686 N
    (c) 6860 J (d) 14 m

8 See page 28

9 160 m

10 (a) 100 m
    (b) 24 m
    (c) 40:24   100   140

11 (a) 0.4 kg m/s
    (b) 6 kg m/s
    (c) 2800 kg m/s

12 (a) 0 kg m/s
    (b) 0 kg m/s
    (c) $m_A v_A - m_B v_B = 0$
    (d) −24 m/s (sign indicates opposite direction to car A)

13 (a) $MV - mv = 0$
    (b) −0.3 m/s (change 30 g to kg; sign shows direction)

14 (a) 10 kg m/s
    (b) 10 kg m/s
    (c) 100 N
    (d) (i) longer time to stop ball
         (ii) smaller force (impulse constant)

15 1080 N

16 (a) larger (b) opposite

17 (a) 1/3 (b) less (c) high

18 (a) 1/3 (b) 3 times/s
    (c) 2.5 m (d) 7.5 m/s

19 (a) force multiplier
    (b) 5
    (c) more

20 (a) 300 metres/megajoule
    (b) 30 miles/gallon
    (c) 25 metres/megajoule, 2.5 miles/gallon
    (d) car 120, train 10
    (e) 48 passengers

Chapters 13 and 14

1 (a) parabola.
   (b) independent.
   (c) its weight.
   (d) the air resistance.

2 (a) parabolic (b) 2 s

3 (a) faster, Bernoulli, curved, force

4 (a) air speed greatest above wings, pressure greatest below wings
   (b) weight of aeroplane
   (c) 4000 N/m^2

5 (a) need air
   (b) jet of gases turns turbine fan – turns turbine axle – turns compressor fan
   (c) air and fuel (kerosene)
   (d) jet of gases backward, thrust forward

6 (a) liquid
   (b) rockets carry their own oxygen
   (c) temperature of liquid oxygen and fuel very low

7 (a) no (b) unbalanced
   (c) centre (d) centripetal.

8 (a) gravitational attraction between Moon and Earth
   (b) attraction between opposite charges
   (c) friction between road and tyres
   (d) tension in the rope

9 (a) 2 N (b) 54 N

10 stone and skater move off at a tangent to the circle

11 (a) (i) satellite falls closer to the ground
         (ii) satellite moves out into space
    (b) satellite moves closer to ground because the gravitational force is now greater than the centripetal force needed to keep it in this orbit.

12 Astronaut and spaceship are both in free fall with the same acceleration towards Earth. So there is no contact force between him and spaceship. Cup and floor of spaceship are both 'falling' towards Earth so cup does not get any nearer the floor.

13 2600 s = 43.3 min (change km into m)

14 (a) Mercury
    (b) Jupiter
    (c) Mars
    (d) Saturn, Jupiter and Uranus
    (e) Pluto
    (f) Mercury

15 (a) distance travelled by light in one year
    (b) 9.5×10^{12} km (change m to km)

16 (a) Neptune
    (b) Venus
    (c) Pluto

17 (a) nuclear fusion
    (b) (ii) $14 \times 10^6\,°C$
    (c) three of: visible light, IR, UV or radio

18 (a) Venus, Mars
    (b) nitrogen
    (c) 30%
    (d) liquid/solid nickel–iron
    (e) rotation of liquid iron sets up electric currents which generate Earth's magnetic field

19 (a) southern, as tilted away from the Sun
    (b) 365 days
    (c) 3×10^7 s
    (d) $v = s/t = (2 \times \pi \times r)/t$
$$= \frac{(2 \times \pi \times 1.5 \times 10^{11})}{(3 \times 10^7)}$$
$$= 3 \times 10^4 \text{ m/s} = 30 \text{ km/s}$$

20 (a) 29.5 days
    (b) Moon also spins on axis once every 29.5 days
    (c) Earth spins on axis once every 24 hours turning different sides towards Sun
    (d) nearly all in line
    (e) neap tides

21 (a) eclipse of the Sun
    (b) See figure 14.3, page 37 (at right)
    (c) umbra
    (d) the Sun, Moon and Earth are not all directly in line every month

Chapters 15, 16, 17 and 18

1 (a) A (b) C (c) D (d) B

2 (a) leaves
   (b) right angles
   (c) towards, away from
   (d) ray

3 (a) 80 cm
   (b) light moves more slowly in
       water than in air

4 (a) violet
   (b) A = deviation of red light
       B = deviation of violet light
   (c) C = dispersion

5 (a) yellow, magenta, cyan
   (b) magenta
   (c) white
   (d) black
   (e) (i) red flower, black leaves,
          black pot
       (ii) yellow flower, green leaves,
          black pot
   (f) green

6 (a) same
   (b) random
   (c) same

7 (a) 50° (b) 50° (c) 80°

8 a and e

9 (a) / (b) ——>>>> (c) {v>×]

10 (a) greater
   (b) total internal reflection

11 (a) 45°
   (b) 90°
   (c) right-angled prisms

12 (a) P = concave, Q = convex,
       R = plane, S = parabolic
   (b) (i) Q = convex
       (ii) S = parabolic

13 (a) thinner
   (b) converges (brings together)
   (c) optical centre
   (d) distance, optical centre

14 (a) A (b) E (c) B or D
   (d) BC or CD (e) C

15 (a) short
   (b) dioptres
   (c) diverging
   (d) metres

16 (a) (i) +4D
       (ii) +2D
       (iii) −10D (must change cm
          into m)
   (b) converging meniscus, +0.50 m

17 (a) E (b) C (c) A (d) F (e) D (f) B

18 (a) real, diminished, inverted
   (b) retina
   (c) flatter, thinner
   (d) increases
   (e) iris
   (f) allow light into eye

19 (a) convex or converging
   (b) virtual, upright, magnified
   (c) increases
   (d) 4

20 (a) real, inverted, magnified
   (b) concentrates light evenly on to
       film
   (c) away from the projector

21 (a) closer to
   (b) four
   (c) blurred

22 (a) short
   (b) on a dull day or when the
       exposure time is short
   (c) f-numbers
   (d) stops

23 (a) 12.5 mm (b) 16

24 (a) long sight
   (b) converging meniscus
   (c) in front of retina
   (d) no

Chapters 19, 20, 21 and 22

1 (a) transverse, longitudinal
   (b) transverse
   (c) transverse
   (d) longitudinal

2 a, c and d

3 (a) travelling
   (b) energy lost or waves spread out
   (c) no
   (d) yes

4 (a) standing
   (b) nodes
   (c) at the antinodes

5 (a) OD, AE, BF, CG, DH
   (b) OP or OQ

6 (a) period
   (b) frequency
   (c) 1/10 = 0.1 s

7 (a) 1.2 m/s

8 (a) straight beam
   (b) dipper
   (c) 30°
   (d) no
   (e) circular waves appear to have
       virtual source 10 cm beyond
       barrier

9 (a) speed
   (b) speed, wavelength, frequency

10 (a) diffraction

   (b) same size as wavelength of wave
   (c) interference

11 (a) See figure 20.6 page 53
   (b) 2a and zero
   (c) A
   (d) anti-phase

12 Q crest + trough destructive zero
   R crest + crest   constructive 2a

13 (a) vibrate
   (b) medium (or material)
   (c) vacuum.

14 (a) longitudinal
   (b) X = compression,
       Y = rarefaction

(c) closer together at X
(d) Y
(e) 0°

15 (a) smaller amplitude
(b) less waves on screen
(c) irregular pattern/constantly changing

16 (a) 4s (double distance)

17 (a) 3000 m (b) 1500 m

18 reverberation

19 (a) A = gamma rays, B = visible light, C = microwaves, D = radio waves

(b) A = gamma rays
(c) ultra violet
(d) each region overlaps the next one

20 (a) energy
(b) interference
(c) not
(d) light or 3×10^8 m/s.
(e) medium

21 (a) gamma or X-rays
(b) visible light
(c) VHF radio waves
(d) microwaves

22 (a) (i) 6×10^{-7} m
(ii) 100 m

(b) (i) 1×10^{20} Hz (change pm to m)
(ii) 3×10^{11} Hz (change mm to m)

23 50 MHz

24 (a) IR (b) UV (c) gamma rays
(d) micro/radio/TV (e) microwaves

25 (a) radio waves
(b) X-rays
(c) X-rays
(d) X or gamma rays
(e) UV

Chapters 23, 24 and 25

1 (a) gains
(b) positively
(c) equal
(d) repulsion

2 a, b and d

3 (a) no (b) yes (c) 2 (d) repel

4 a, d, e and f

5 (a) rods repel, move apart
(b) rods attract, move together
(c) induction leads to attraction between the rods

6 (a) see figure 23.4 page 61
(b) electrons flow to ground
(c) positive

7 (a) dust particles, earthed metal plates
(b) needle in tip of gun negatively charged to high voltage

8 (a) I (b) Q (c) V (d) A (e) C (f) V (g) W

9 (a) 0.5 A (change min to s)
(b) 30 A (min to s)

10 (a) 1800 C (change min to s)
(b) 2880 C (h to s)

11 (a) cell, lamp, switch
(b) closed
(c) see figure 24.1 page 62
(d) from − to + terminals of battery round the circuit.

12 (a) see figure 24.3 page 62
(b) 0.2 A

13 (a) charged
(b) charge
(c) ammeter, series.
(d) equal

14 (a) amps (b) coulombs (c) volts
(d) joules

15 (a) voltage or potential difference
(b) parallel
(c) battery.

16 (a) 2 J/C (b) 5 J/C (c) 240 J/C

17 (a) 240 V (b) 12 V

18 (a) 72 J (b) 5 J

19 (a) flow of charge or current
(b) double
(c) ohm
(d) V/I

20 (a) 8 V (b) 6 V

21 (a) 0.05 A (b) 8 A

22 (a) 4 Ω (b) 100 Ω

23 (a) Ohm's, temperature, proportional
(b) ohmic

24 (a) characteristic
(b) filament lamp
(c) the increases in current get smaller
(d) increases

25 (a) diode or rectifier
(b) forward bias

26 (a) 200 Ω (b) 800 Ω (c) 10 Ω
(d) 2.4 Ω

27 (a) 6 Ω and 30 Ω (b) 5 Ω (c) 10 Ω
(d) 2 Ω and 8 Ω (e) 10 Ω (f) 5 Ω

28 (a) 0.2 A (b) 2 V (c) 15 Ω

Chapters 26, 27, 28 and 29

1 (a) north, south
   (b) unlike
   (c) soft
   (d) yes
   (e) repulsion

2 (a) see figure 26.2a page 68
   (b) north to south
   (c) pattern same as in figure 26.2b page 68, but lines in opposite directions
   (d) neutral point where the field is zero

3 (a) (i) drop off
         (ii) stay attached
   (b) nails

4 (a) stroking
   (b) keeper

5 (a) towards Earth's North pole
   (b) Earth's North Pole is the south pole of the Earth's magnetic field
   (c) declination

6 (a) magnetic, circular field
   (b) Maxwell's screw rule
   (c) close, parallel, evenly spaced

7 field pattern = concentric circles in A: anticlockwise, in B: clockwise

8 (a) high, direct current
   (b) alternating

9 insulated: so wires don't touch and short circuit; copper wire: to conduct current; soft-iron: temporarily magnetised (when I flows); switch: to turn current on/off; battery: provides direct current to magnetise core.

10 (a) 500 J
    (b) 120 000 J or 120 kJ (change min to s)
    (c) 3 600 000 J or 3.6 MJ (change h to s)

11 (a) 192 W (b) 0.75 W

12 (a) wire, low, fuse.
    (b) Qu. 10: (a) 2 A (b) 3 A (c) 5 A
         Qu. 11: (a) 1 A (b) 1 A

13 b, d and e

14 c, e and f

15 (a) live
    (b) faulty, cables, overloaded
    (c) safety

16 (a) blue = Neutral
         brown = Live
         green/yellow = Earth
    (b) A = neutral
         B = earth
         C = live
    (c) 2 A or 3 A

17 (a) parallel
    (b) 0.24 kWh
    (c) 1.2p

18 (a) 5472 (ignore last dial)
    (b) 430

19 (a) left to right through magnet
    (b) north to south (top to bottom)
    (c) down into paper
    (d) reverse battery or magnet

20 (a) (i) up (out of paper)
         (ii) down (into paper)
    (b) clockwise (as viewed from AD)
    (c) stronger magnets or more turns of wire

21 (a) direct current
    (b) more powerful magnet, longer pointer or weaker spring
    (c) rotor
    (d) commutator
    (e) reverses current direction to keep coil turning

22 (a) current I flows and ammeter deflection is in one direction
    (b) I flows opposite way and ammeter deflection is opposite way
    (c) bigger I, bigger deflection

23 (a) induced, magnetic
    (b) electromagnetic
    (c) bigger
    (d) currents

Chapters 30, 31, 32 and 33

1 (a) electron
   (b) cathode
   (c) light
   (d) Y-plates

2 (a) negatively
   (b) -1.6×10^{-19} C
   (c) smaller

3 (a) emission of electrons from heated metal or metal oxide
   (b) (i) and (iii)

4 (a) coils around tube
   (b) 625
   (c) to reduce flickering effects
   (d) red, green, blue

5 (a) 6 V
   (b) spot moves down to 3 divs below centre
   (c) alternating amplitude = ± 6 V
   (d) time-base switched on
   (e) 12 V

6 1 W

7 (a) $22\,000\,\Omega \pm 5\% = 22\,k\Omega \pm 5\%$
   (b) $100\,\Omega \pm 10\%$

8 (a) $4.7\,k\Omega \pm 5\%$
   (b) $0.33\,M\Omega \pm 20\%$
   (c) $56\,\Omega \pm 10\%$
   (d) $10\,k\Omega \pm 5\%$

9 (a) (i) blue grey red silver
         (ii) 6K8K
   (b) (i) yellow violet brown gold
         (ii) K47J
   (c) (i) grey red yellow (no band)
         (ii) M82M

10 (a) current
    (b) voltage
    (c) voltage (or potential)

11 (a) 6 V
    (b) 4 V
    (c) 8 V
    (d) 9 V
    (e) 8.4 V

12 (a) decreases
    (b) temperature
    (c) forward
    (d) rectification
    (e) diode or rectifier
    (f) see page 81, figures: (i) 31.3
         (ii) 31.6 (iii) 31.7

13 c, d, f and g

14 (a) 3
    (b) b = base, c = collector,
         e = emitter
    (c) I_b = base current, I_c = collector
         current

15 (a) 6 mA

    (b) 0.002 A = 2 mA
    (c) 100 μA
    (d) 50 mA

16 (a) off (b) on

17 (a) (i) JK
         (ii) MN
    (b) (i) LDR
         (ii) thermistor
    (c) (i) LDR in dark
         (ii) thermistor cold
    (d) reverse positions of sensor and R
    (e) R

18 (a) relay
    (b) relay to switch large currents
         needed to operate heater
    (c) diode protects transistor from
         high voltages produced across
         relay coil when current stops
         flowing

19 (a) voltage, current
    (b) continuously.

20 (a) 100 000

    (b) very high
    (c) yes
    (d) inputs: + = non-inverting,
         − = inverting

21 (a) positive
    (b) reduces distortion
    (c) gain constant for a wide range of
         frequencies

22 (a) inverting input
    (b) inverting
    (c) (i) 10 kΩ
         (ii) 20 kΩ
    (d) gain = $-R_f/R_{in}$ = -2
    (e) ∓ 10 V
    (f) 10 × larger and inverted
    (g) ∓ 15 V (saturation)

23 (a) saturation = clipping effect,
         output voltage prevented from
         reaching its full amplified value
         because it cannot exceed the
         power supply values.
    (b) frequency response
    (c) voltage gain
    (d) bandwidth
    (e) reduces

Chapters 34, 35 and 36

1   (a) number
    (b) binary
    (c) change

2   see figure 34.1 page 88

3   (a) NOT
    (b) NOR
    (c) NAND
    (d) NOT
    (e) AND

4   (a) NAND gate
    (b) OR gate
    (c) AND gate
    (d) NOR gate (truth tables on page
         88)

5   (a)

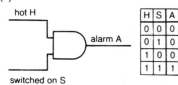

H	S	A
0	0	0
0	1	0
1	0	0
1	1	1

(b)

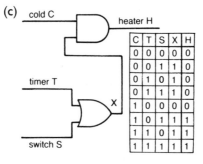

D	S	L
0	0	0
0	1	1
1	0	1
1	1	1

(c)

C	T	S	X	H
0	0	0	0	0
0	0	1	1	0
0	1	0	1	0
0	1	1	1	0
1	0	0	0	0
1	0	1	1	1
1	1	0	1	1
1	1	1	1	1

6   (a) light-emitting diode
    (b) see figure 34.2 page 88
    (c) to protect the LED by limiting
         the current
    (d) small

7   (a) in series

    (b) 3.3 V
    (c) 10 mA
    (d) 330 Ω

8   (a) memory
    (b) output, reset.

9   (a) see figure 34.5 page 89 but
         change positions of LDR and
         resistor
    (b) see figure 34.5 page 89 but
         change the resistor for a
         thermistor and the LDR for a
         resistor

10 (a) noise
    (b) power of signal
    (c) modulating

11 (a) amplitude modulation
    (b) frequency modulation
    (c) much higher
    (d) FM
    (e) see page 90 figures: (i) 35.1
         (ii) 35.2

12 (a) pulse code modulation
   (b) binary-coded pulses

13 (a) AF
   (b) tuner
   (c) separate AF from RF
   (d) AF

14 (a) FM
   (b) sound, vision and sync. separator
   (c) by using line and field time-bases
      at the right moments

15 (a) concave
   (b) high
   (c) Earth's surface curved
   (d) to amplify signals because they
      get weaker as they travel

16 (a) decrease 1/4 times
   (b) decrease 1/16 times

17 (a) geostationary or
      geosynchronous
   (b) PCM

18 (a) element
   (b) compound
   (c) molecules

19 (a) solid, liquid, gas
   (b) (i) gas (ii) solid (iii) liquid
   (c) (i) solid (ii) liquid (iii) solid
   (d) gas

20 (a) rubber
   (b) cast iron
   (c) plasticine

(d) glass
(e) copper

21 (a) ductile
   (b) plastic
   (c) strong
   (d) stiff

22 (a) 40 cm
   (b) 5 cm
   (c) 40 cm
   (d) Hooke's law

23 (a) (i) 0.7 N (ii) 1.0 N
   (b) (i) 2.2 cm (ii) 0.6 cm
   (c) 1.8 cm

24 see figure 36.6 page 93

Chapters 37, 38, 39, and 40

1 (a) energy
   (b) temperature
   (c) thermometer

2 (a) steam point, 100 °C
   (b) ice point, 0 °C

3 (a) absolute zero
   (b) (i) 0 K (ii) −273 °C

4 (a) 1083 °C
   (b) 4 K

5 (a) clinical
   (b) thermistor
   (c) thermocouple
   (d) alcohol-in-glass

6 (a) increases
   (b) expand
   (c) to expand by differing amounts
      so it bends

7 (a) capacity
   (b) specific

8 (a) 4200 J/K
   (b) 21 000 J/K

9 (a) 100 °C
   (b) 504 000 J

10 (a) liquid to vapour (or gas)
   (b) vapour to liquid

(c) liquid to solid
(d) solid to liquid
(e) vapour to solid (or solid to
    vapour) without being a liquid

11 (a) flat part BC
   (b) 70 °C
   (c) latent heat

12 (a) stays constant
   (b) 350 J (change g to kg)

13 (a) at the surface
   (b) faster (so have enough energy to
      escape)
   (c) falls
   (d) to increase evaporation (by
      blowing away molecules so they
      cannot return to the liquid)

14 (a) vapour forms bubbles – rise to
      surface and burst
   (b) stays constant
   (c) 4.6×10^6 J

15 (a) kinetic theory
   (b) kinetic
   (c) internal

16 (a) illuminated smoke particles
   (b) randomly, jerkily, in all directions
   (c) moving air molecules bump into
      smoke particles
   (d) Brownian motion

17 (a) randomly, slowly, lots of
      collisions with each other
   (b) pale blue
   (c) diffusion

18 (a) (i) doubles (ii) increases three
         times
   (b) 45 litres

19 (a) (i) halves (ii) doubles
   (b) 0.08 m³

20 (a) (i) doubles (ii) reduces by three
            times
   (b) 150 kPa

21 (a) 8 m³ (change °C to K)
   (b) 1 m³

22 (a) conduction
   (b) by molecules vibrating and so
      colliding and bumping each other
   (c) molecules very far apart so not
      many collisions between them
   (d) metals have free electrons which
      move by diffusion and transfer
      energy by collisions with the
      molecules

23 (a) wood
   (b) single
   (c) thicker

24 convection

25 (a) conduction
   (b) hot water expands so is less dense, so rises; cold water is more dense so sinks

26 (a) infrared
   (b) light $= 3 \times 10^8$ m/s
   (c) dull black
   (d) shiny light coloured

27 (a) glass
   (b) silver
   (c) vacuum
   (d) cork or plastic
   (e) c and d

Chapters 41, 42 and 43

1 (a) cells
  (b) oxygen

2 (a) IR radiation
  (b) conduction
  (c) evaporation
  (d) evaporation
  (e) conduction

3 (a) industry
  (b) 25%
  (c) heating
  (d) 5%

4 (a) kilowatt-hour
  (b) therm
  (c) tonne
  (d) litre

5 (a) 2500
  (b) 9000 MJ $= 9 \times 10^9$ J

6 33 000 MJ $= 33 \times 10^9$ J

7 b, d and f

8 (a) (i) coal
      (ii) electricity
      (iii) coal
      (iv) electricity
  (b) (i) 275 MJ
      (ii) 190 MJ
  (c) gas

9 (a) wind
  (b) sunlight
  (c) tidal
  (d) radiant (IR)
  (e) coal, oil or natural gas

10 (a) reflected or absorbed and radiated back
   (b) (iii)

11 a, c and d

12 b, e, f and g

13 (a) wind
   (b) mechanical kinetic
   (c) (iv)

14 (a) Incoming tides raise water level behind a dam. As tide goes out, the trapped water is released through holes in the dam and drives turbines
   (b) gravitational potential

15 (a) heat from hot rocks deep in the Earth
   (b) cold water sent down returns as hot water or steam
   (c) gases including radioactive radon

16 (a) 2000 MW $= 2 \times 10^9$ W (change kW and km to W and m)
   (b) 1000 MW ($= 1 \times 10^9$ W) $=$ 1000 MJ/s
   (c) their energy has been extracted

17 (a) expensive and average only 10 W/m^2 of cells
   (b) improves the absorption of radiation

18 (a) heat energy
   (b) (i) releases heat
       (ii) absorbs heat
   (c) evaporator
   (d) inside

19 (a) stored (chemical)
   (b) electrical
   (c) kinetic
   (d) stored (elastic potential energy)
   (e) sound
   (f) stored (gravitational potential energy)

20 For example:
   (a) microphone

   (b) battery
   (c) light bulb
   (d) electric motor

21 (a) energy changes
   (b) form of energy
   (c) reversible

22 chemical energy in candle to heat and light

23 (a) A = chemical energy in battery, B = electrical, C = kinetic energy of moving parts, D = potential energy of raised load
   (b) heat losses: E = in wires carrying current, F = in the bearings of rotating parts

24 (a) coil, oil or uranium
   (b) primary–heat–potential–kinetic–electrical
   (c) heat

25 (a) method of saving surplus electrical energy when demand is low
   (b) uphill
   (c) gravitational potential energy

26 (a) 2 MJ $= 2 \times 10^6$ J
   (b) 2 MJ/s
   (c) 1.6 MW

27 (a) energy can neither be created nor destroyed
   (b) heat energy

Chapters 44, 45, 46, 47 and 48

1 (a) network of cables linking power stations throughout Britain
   (b) waste heat, voltage drop
   (c) thick cable, high voltage
   (d) power loss $= I^2R$

2 (a) 20 A (power $= VI$ and change kW to W)
   (b) 2000 W (I^2R)
   (c) 100 V (power loss/I)

3 33 kV

4 (a) alpha-particle scattering
   (b) (i) very small (about 1/10 000 of diameter of atom)
       (ii) positive
       (iii) in orbits or shells around the nucleus

5 (a) protons, neutrons, electrons
   (b) protons and neutrons
   (c) neutron
   (d) electron
   (e) proton

6 (a) Z
   (b) N
   (c) A

7 (a) $Z = 4$, $N = 6$
   (b) $Z = 11$, $N = 12$
   (c) $Z = 14$, $N = 17$
   (d) $Z = 56$, $N = 84$

8 (a) Both isotopes have six protons and six electrons. $^{12}_{6}$C has six neutrons but $^{14}_{6}$C has eight neutrons
   (b) different numbers of neutrons

9 (a) (i) helium nucleus (two protons and two neutrons)
       (ii) electron
       (iii) electromagnetic radiation, high frequency
   (b) (i) β (ii) γ (iii) γ (iv) α and β (v) α

10 (a) alpha particles collide with and so ionise the air molecules
   (b) ions
   (c) light is reflected from the drops
   (d) straight, thick, about 3 to 4 cm long

11 (a) γ (b) α (c) α (d) β, γ

12 (a) Geiger–Müller tube and pulse counter
   (b) (i) 1/4 (ii) 1/8

13 (a) 25%
   (b) 20 years
   (c) background count rate subtracted to give count rate for lead alone
   (d) see page 117

15 (a) 8 g
   (b) 4 g
   (c) 1 g
   (d) 0.5 g

16 (a) $^{227}_{90}$Th → $^{223}_{88}$Ra + $^{4}_{2}$α
   (b) $^{16}_{7}$N → $^{16}_{8}$O + $^{0}_{-1}$β

17 (a) fission
   (b) neutrons
   (c) neutrons
   (d) energy
   (e) critical

18 (a) slows down the neutrons
   (b) allows more neutrons to be absorbed by uranium-235

   (c) graphite or pressurised water
   (d) control
   (e) lowered into reactor to absorb neutrons

19 (a) A = kinetic energy of neutrons, B = heat in reactor core
   (b) gas or water

20 (a) breeder
   (b) plutonium
   (c) the neutrons are moving fast
   (d) no, because the neutrons don't have to be slowed down to thermal neutrons (fast neutrons are used)

21 (a) fusion
   (b) deuterium and lithium

22 (a) ions
   (b) sieverts
   (c) (i) and (iii)

23 all

24 (a) electrocardiograph
   (b) electrical voltage generated in heart to contract muscles
   (c) about 1 mV
   (d) 86 beats/min (time period = 0.7 s, pulse rate per min = 60/0.7)

25 (a) ultrasound
   (b) thermography
   (c) optical fibre endoscopy
   (d) X-rays

26 (a) pacemaker
   (b) a laser

Answers to structured and exam-type questions

The following answers are entirely the responsibility of the authors, and have not been provided by the examination boards.
(*Note:* MEG do not allow answers to the questions marked * to be given.)

1. (a) less
   (b) (i) cork, ice
       (ii) cork
       (iii) concrete
       (iv) concrete, perspex, ice

2. (a) 6mm
   (b) 36mm^2
   (c) 31 squares
   (d) total area = area of 1 square × number of squares = 36mm^2 × 31 = 1116mm^2

3. (a) contact force or push on tyre from road
   (b) 400N
   (c) 1400kg
   (d) speed increases because of excess force pushing forward

4. (a) graph
   (b) decrease
   (c) (i) 6.4N/kg
       (ii) 1.8N/kg
   (d) g = 4.9N/kg so h about 2600km (actually h = 2650km)
   (e) (i) 9800N
       (ii) 1000kg
       (iii) 9200N

5. (a) 0.25Nm
   (b) 0.25 = 0.2W_{bb}, W_{bb} = 1.25N
   (c) 1.25g
   (d) end A tips *up*, weight of new ball bearings < weight of old ball bearings

6. (a) C of G = point in space directly below pencil point
   (b) stable because C of G lies below the pivot
   (c) C of G in middle, half way along pencil
   (d) unstable as C of G now above pivot

7. (a) force up = upthrust U, force down = weight W
   (b) sinks as upthrust is reduced and weight now greater than upthrust

   (c) pressure on her increases with vertical depth of water above her

8. (a) oil
   (b) (i) force on brake pedal – magnified by lever system – pressure transmitted through brake fluid (which is incompressible) to piston – force on brakes
       (ii) as brake pedal moves – increase in pressure compresses air bubble – but pressure in system (oil + air) does not reach a high enough value to operate brakes
       (iii) pressure causes more fluid to leak – less pressure transmitted to piston – less force on brakes

9. (a) decrease
   (b) (i) 100kPa
       (ii) about 42kPa
   (c) 40km
   (d) air pressure inside aeroplane much greater than outside so air leaves the aeroplane rapidly!

10. (a) see figure 5.3 page 12
    (b) pressure is greatest outside
    (c) to prevent air entering behind the sucker and so raising the pressure
    (d) clean, moist, smooth surfaces give a better seal

11. (a) (i) 0.06W (power = force × speed and speed = height × 3 rev/s)
        (ii) 0.12W
    (b) (i) A: 0.15W, B: 0.3W, C: 0.42W (output = input × efficiency)
        (ii) eff = A: 0.06, B: 0.12, C: 0.17
             A: power too low, speed too low
             B: power matches, select B
             C: too much power, too quick/wasteful

12. (a) 3600W
    (b) 80%
    (c) heat due to friction of bearings
    (d) 8640J/min (change s to min)
    (e) lubricate with oil

13. (a) lever ◣
    (b) 1:16
    (c) 1/4cm
    (d) 20 times
    (e) handle: 30π = 94mm, blade: 6π = 19mm
    (f) 5:1

14. (a) 3
    (b) changes direction of the force
    (c) pulleys allow the length of wire to change with temperature while preventing sagging and maintaining constant tension

152

15 (a) 2 m/s
   (b) 600 s (change min to s)
   (c) 1 m/s
   (d) back at the car park
   (e) 4800 m

16 (a) graph
   (b) 10 m/s
   (c) 20 m/s
   (d) 30 s
   (e) 0.5 m/s^2 ((20–10)/2)
   (f) 1 m/s^2 ((70–50)/20)
   (g) 200 m

17 (a) 800 m/s^2 (0.8/0.001)
   (b) (i) 0.067 s ($v=u+a\,t$)
       (ii) 0.027 m ($s=(u+v)t/2$)
   (c) air resistance also acts downwards

18 (a) 162 m ($s=(u+v)t/2$)
   (b) 12 s ($v=s/t$ and $s = 810 - 162$)
   (c) 18 s
   (d) air resistance increases with speed until resistance = weight
   (e) less air offers less resistance

19 *

20 (a) 600 N
   (b) 600 N
   (c) (i) force needs to be greater than 600 N to accelerate person
       (ii) 720 N ($F_1 + F_2 = 600 + m\,a = 600 + 120\,\text{N}$)

21 (a) graph
   (b) straight line through origin or braking distance directly proportional to (initial speed)^2
   (c) 400 (m/s)^2
   (d) 24 m
   (e) doubles
   (f) (i) directly proportional ($E_k=\tfrac{1}{2}mv^2$)
       (ii) directly proportional as braking distance $\propto E_k \propto v^2$

22 (a) 20 m/s
   (b) 0 m/s
   (c) 20 m/s
   (d) 1500 kg m/s
   (e) 30 000 N = 30 kN ($Ft = m\Delta v$)
   (f) doubles time, halves force
   (g) impulse = $m\Delta v$ = constant so no change in impulse

23 (a) to make the bit turn faster, or speed or distance multiplier
   (b) 8 rev/s
   (c) more teeth on big wheel, less teeth on small wheel (i.e. increase gear ratio)

24 (a)

	car A	car B	car C	
(i)	4 m/s^2	6 m/s^2	5 m/s^2	($v = u + at$)
(ii)	3200 N	4800 N	4000 N	($F = ma$)
(iii)	1.44 MJ	1.44 MJ	1.44 MJ	($E_k = \tfrac{1}{2}mv^2$)
(iv)	96 kW	144 kW	120 kW	($P = E_k/t$)

   (b) B
   (c) B
   (d) B
   (e) A

25 (a) 2000 kg
   (b) 20 000 kg m/s
   (c) −20 000 kg m/s
   (d) −10 m/s

26 (a) orbits so that it hovers over fixed position on Earth
   (b) 24 hours
   (c) 540 N
   (d) 540 N
   (e) 3286 = 3.3 km/s
   (f) F = four times bigger, v increased to 4.6 km/s ($F = mg = mv^2/r$)

27 (a) Neptune
   (b) Uranus
   (c) Jupiter
   (d) Saturn, Uranus
   (e) Mars, Pluto
   (f) Saturn, Pluto
   (g) radiation from Sun gets weaker in proportion to the square of the distance from the Sun
   (h) greenhouse effect, radiation penetrates clouds and heats rocks, radiation from rocks absorbed by clouds so raising temperature of planet.

28 (a) Earth
   (b) (i) total
       (ii) partial
   (c) 3.6×10^6 m/h ($v = s/t$)

29 (a) and (b) see figures 15.1 and 15.2 page 40 (draw a normal where rays leave water)
   (c) slows

30

	1	2	3	4	5	6
(a)	red	blue	all	green	none	Y, red, Gr
(b)	red	black	yellow	green	black	yellow
(c)	black	black	yellow	black	black	yellow

31 (a) R
   (b) 2
   (c) 30°
   (d) 30°
   (e) equal

32 (a) inner part shiny, bright; outer part dull, matt
  (b) inner
  (c) irregular or diffuse

33 (a) see figure 18.2 page 46
  (b) (i) move lens in/out
   (ii) towards screen
   (iii) invert slide
  (c) 0.80 m (100 × 8 mm)

34 (a) (i) iris
   (ii) ciliary muscles
   (iii) lens + cornea
  (b) see figure 17.8 page 45
  (c) nine times area
  (d) stereo vision gives a sense of distance

35 (a) see figure 18.1 page 46
  (b) virtual, upright, magnified
  (c) magnifying glass/simple microscope

36 (a) see figure 18.3 page 46
  (b) converging or convex
  (c) see figure 18.4 page 47
  (d) smaller, inverted
  (e) in lens camera: image sharper and 'collects' more rays, so image brighter

37 (a) 4000 m
  (b) 200 m
  (c) decreases
  (d) decreases

38 (a) standing or stationary, reflect, overlap
  (b) 1.5
  (c) 2 m
  (d) 20 Hz
  (e) decrease length, increase tension

39 (a) and (b) see figures 20.1 and 20.2, page 52
  (c) (i) 1.5 cm
   (ii) 15 cm/s = 0.15 m/s

40 (a) X=large amplitude waves, Y=sound, Z=dark area
  (b) constructive interference
  (c) constructive interference
  (d) destructive interference – see figure 20.6 page 53

41 sound totally internally reflected – see figure 16.5 page 43 – same for sound

42 (a) (i) see figure 21.2 page 54 but replace loudspeaker with tuning fork
   (ii) 1.36 m
  (b) (i) see figure 21.3B page 54 + period marked (e.g. peak to next peak)
   (ii) amplitude increases – see figure 21.3A

(iii) halved time period so twice as many waves on screen

43 (a)

radio	1	2	3	4
λ (m)	285	330	370	300
f (kHz)	1052	910	810	1000

  (b) $\lambda f = 3 \times 10^8$ m/s in each case, so $v = \lambda f$
  (c) $\lambda = v/f = 3 \times 10^8/(1100 \times 10^3) = 273$ m

44 (a) walls made of metal
  (b) remember angle of incidence = angle of reflection
  (c) to ensure all parts of food cook equally
  (d) gets smaller, wave absorbed by water
  (e) e.g. energy only goes to food, and reaches all parts of food, does not have to be conducted, no warm-up time of oven
  (f) wavelength smaller ($c = f\lambda$, c smaller, f fixed so λ smaller)

45 (a) + on top, − on bottom
  (b) some free electrons move to bottom leaving deficit of electrons on top
  (c) return to original state – disappears

46 (a) positive
  (b) like charges repel
  (c) cheque negative and powder now negative so repel

47 L_1 and L_2 in series, L_3 in parallel

48 (a) 9 V
  (b) 4.5 V
  (c) 4.5 V
  (d) 9 V

49 (a) level of fuel alters float – alters value of variable R – alters I flowing in circuit and reading on ammeter (fuel gauge)
  (b) add fuel to the tank a litre at a time and mark off the positions of the pointer on the gauge

50 (a) 7 Ω
  (b) 3 Ω
  (c) 10 Ω
  (d) 0.8 A
  (e) 5.6 V
  (f) 2.4 V
  (g) (i) 0.6 A
   (ii) 0.2 A

51 (a) induced magnetism
  (b) south
  (c) repulsion as ends of needles far from magnet are induced north poles

52 (a) left end = north, right end = south
(b) arrows follow direction of field lines, see figure 26.2 page 68

53 *

54 (a) 3 A
(b) 5 A
(c) 144 W
(d) 20%
(e) 43 200 J

55

	cooker	heater	drier
(a)	8 kW h	2 kW h	0.32 kW h
(b)	40p	10p	1.6p

(c) 51.6p = 52p to nearest pence

56 (a) see figure 28.4 page 73
(b) 30 A in fuse box
(c) (i) separate cable from the fuse box to cooker
(ii) in ring main cable, the loop provides two parallel wires connecting each socket to the fuse box, so the current is shared by the two wires, so wires can be thinner: the single cable to cooker carries all current, so is thicker
(iii) earth wire, if fault allows the live wire to touch earthed case of appliance then earth wire (low resistance) allows a high current to flow and blow the fuse

57 (a) magnet
(b) induced magnetism, poles alternate
(c) coil
(d) turn wheel faster
(e) a.c. as induced magnetism alternating
(f) no movement of magnet so no reversal of field through core and coil

58 (a) soft-iron
(b) can be temporarily magnetised
(c) X
(d) use larger current or more turns of wire
(e) increases strength of magnetism of recording head and so induced magnetism in tape

59 (a) 1 V/cm
(b) 4 V
(c) 0.5 ms/cm
(d) 5 ms
(e) 200 Hz
(f) 5 cm from peak-to-peak horizontally, 4 cm peak-to-peak vertically

60 (a) off
(b) 2.8 V (2 V/div × 1.4)
(c) 1.4 V
(d) $1.4 = 1.4 \times V_{rms}$, $V_{rms} = 1$ V

61 (a) 4 kΩ
(b) 6 V
(c) (i) 2 kΩ
(ii) 40 °C
(d) (i) 8 kΩ
(ii) 6 °C
(e) non-linear scale, e.g. 4 V = 40 °C, 6 V = 22 °C, 8 V = 6 °C, note low voltage means high temperature

62 (a), (b) and (c) see figure 31.8 page 81

63 (a) (i) sound – mechanical – electrical
(ii) electrical – mechanical – sound
(b) varying I_b causes varying I_c and I_c, being larger than I_b, operates loudspeaker

64 (a) 0.6 V
(b) takes time to charge capacitor until p.d. across it exceeds 0.6 V and turns on transistor
(c) increase resistor value, increase capacitor value
(d) shorts out C so discharges C so no p.d. across C so transistor switches off and bell stops ringing

65 (a) (i) −2.5 V
(ii) +10 V
(b) ±15 V
(c) ∓3 V
(d) inverted a.c. signal of amplitude ∓10 V

66 (a) −10
(b) horizontal line at gain = 10
(c) 10 kHz
(d) 100
(e) amplifies all audible frequencies by same amount

67 *

68 (a)

A	B	X	Y	LED
0	0	1	0	off
0	1	0	1	on
1	0	0	1	on
1	1	0	1	on

(b) LED lit if A or B closed
(c) OR gate
(d) latches so LED stays on once A or B closed even if opened afterwards
(e) see figure 34.5 page 89

69 (a) A, microphone; B, amplifier; E, transmitting dish and aerial; F, satellite, relays and amplifies microwave signal; G, receiving dish, concentrates microwave signal on to aerial; H, separates pulse code from carrier wave; I, decoder, converts PCM digital signal into electrical analogue signal; J, amplifier, amplifies the analogue signal; K, converts electrical signal into sound

(b) W, sound waves; X, electrical analogue; Y, digital pulses; Z, PCM microwave link

70 (a) speech needs: telephone microphone transmitter and earpiece receiver, computer data needs: modem at both ends in addition to above
(b) coaxial cables, optical fibres, microwave links
(c) e.g. amount of info. available, speed of communication, messages can be sent anywhere instantly

71 (a) rule placed vertically, zero at bottom, reading of load position: eye in line, several readings, a zero error at bottom
(b) (i) and (ii) graph and line
(iii) about 1.2 to 1.4 N
(iv) about 177 to 179 mm
(v) about 0.090 to 0.095 N

72 (a) see figure 36.3 page 93
(b) no, not a straight line through origin
(c) (i) graph has similar shape, but extension halved for a given force
(ii) same as graph in (i)
(d) (i) length halved so extension values halved
(ii) load shared between two bands, so half force only on each, so extension values halved.

73 (a) plastic
(b) X and Y expand by different amounts when hot (X expands more than Y) and as they are fixed together the strip bends
(c) contacts move apart, heater no longer connected to power supply, no current flows
(d) decrease as contacts broken at lower temperature

74 (a) (i) X (ii) Y (iii) Z
(b) heat transferred to surroundings, increases as temperature rises
(c) (i) 1.7 °C/min (ii) 4760 J/s (iii) 4760 W
d) lag tank with insulating material so energy from heater heats water and not the room

75 (a) 7560 J $(mc(t_1 - t_2))$
(b) 34 000 J (ml)
(c) 41 560 J

76 (a) water evaporating from wet hair has a cooling effect on head
(b) water vapour in breath condenses on cold mirror

77 (a) and (b) graph and curve
(c) 55 to 57 cm^3
(d) effect of atmospheric pressure

78 (a) molecules: rapid motion, many with wide range of speeds, strike walls repeatedly so average

force is constant. Pressure = force/area also constant
(b) intermolecular forces hold molecules in place, vibration increases with temperature, they break free when energy exceeds a critical value

79 *

80 (a) (i) 35% (ii) 10%
(b) 20%, 20%, 10%, 5%, 5%
(c) 60%
(d) 240 MJ

81 *

82 (a) (i) food, heat
(ii) working, heat
(iii) burns, impacts
(b) sweating – cooling effect of evaporation
(c) hairs rise trapping air – insulator

83 (a) traps solar radiation (greenhouse effect), and reduces heat loss by convection
(b) black, helps improve absorption of radiation
(c) raise hot water tank above level of solar panel so have a head of water, hot water rises and cold water sinks
(d) angle affects amount of radiation received from Sun
(e) insulating material e.g. expanded polystyrene board

84 (a) March
(b) Jan, Feb, Mar, Apr, May, Nov, Dec
(c) (i) 31 units
(ii) power = power/m^2 × area
so area = 1000 units/(50 units/m^2) = 20 m^2

85 *

86 (a) heat, sound, electricity
(b) friction, vibration, alternator

87 (a) 250 A $(P = VI)$
(b) 25 000 V = 25 kV $(V = IR)$
(c) 6 250 000 W = 6.25 MW $(P = VI)$
(d) 6.25%
(e) heat in transmission lines
(f) higher voltage gives lower current and less heat as heat depends on I^2
(g) lower I allows thinner cables so less material
(h) (i) easy maintenance
(ii) environmental issue

88 (a) very tiny and much smaller than the atom
(b) protons and neutrons
(c) 196
(d) very very tiny, about 196 times smaller than the nucleus

(e) electrons

(f) 79

89 (a) (i) number of disintegrations/second

(ii) nuclides with different numbers of neutrons, but same numbers of protons

(b) take several readings for count rate halving e.g.

half-lives:	1	2	3
activity:	40	20	10
time in s:	54	110	166
time-interval:	54	56	56

average half-life = 55 s

(c) (i) alpha particles, thick, short and straight tracks, dense ionisation, tracks of different lengths indicate different energies

(ii) collision with a nucleus of an air molecule which causes alpha particle to change direction

(d) $^{238}_{92}X = ^{234}_{90}Y + ^{4}_{2}\alpha$

90 (a) beta; reject alpha because absorbed by box, gamma because not absorbed by soap

(b) source with a very short half-life would need replacing too often

(c) source D

(d) count rate rises if soap level too low because radiation not absorbed by soap

(e) enclose source in lead housing, restrict access to source

91 (a) consider: opposing claims that there is/is not evidence of illness among workers, risks of accidents

(b) consider: risks of leaks/explosions, hazards of transporting fuel and waste products

(c) yes, e.g. radioactive leaks into atmosphere and rivers/sea, water discharged into surrounding rivers/sea is warm

(d) transported to reprocessing plants where unspent fuel is recovered and waste is stored while activity decays

92 (a) chain reaction slows and produces less heat as more neutrons absorbed

(b) chain reaction speeds up and gets out of control leading to melting/explosion of the core, as number of neutrons increases

(c) chain reaction very slow as fast neutrons cause few fissions unless slowed by a moderator for easy capture by uranium nuclei

(d) core melts as it overheats

93 (a) α, β, γ and X as they have energy and cause ionisation

(b) and (c) see pages 120 and 121

94 (a) and (b) see pages 120 and 121

Symbols and Units

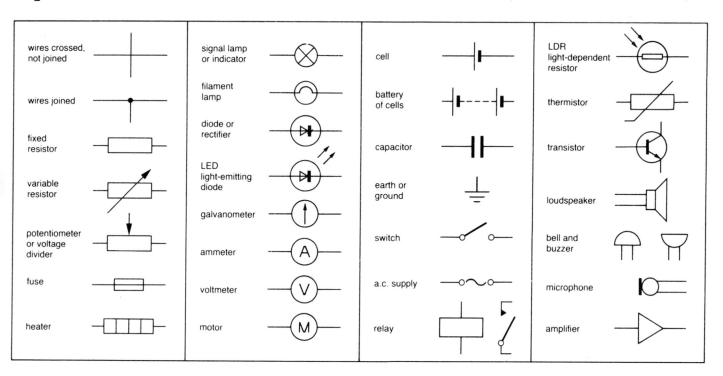

Electrical symbols.

Symbol	Quantity	Unit	Unit Symbol
a	Acceleration	Metre per second squared	m/s^2
a	Amplitude	Metre	m
A	Area	Metre squared	m^2
A	Nucleon number		
A	Voltage gain or amplification		
c	Critical angle	Degree	$\circ$
c	Speed of waves (of light)	Metre per second	m/s
c	Specific heat capacity	Joule per kilogram kelvin	$J/kg\,K$
C	Count rate	Counts per second	$/s$
C	Heat capacity	Joule per kelvin	J/K
d	Density	Kilogram per metre cubed	kg/m^3
d	Distance	Metre	m
E	Energy	Joule (= newton metre)	J
E_p	Potential energy	Joule	J
E_k	Kinetic energy	Joule	J
f	Focal length of a lens	Metre	m
f	Frequency	Hertz (= per second)	Hz
F	force	Newton (= kg metre/s^2)	N
g	Acceleration of free fall	Metre per second squared	m/s^2
g	Gravitational field strength	Newton per kilogram	N/kg
h	Height	Metre	m
i	Angle of incidence	Degree	$\circ$
I	Electric current	Ampere (amp)	A
l	Specific latent heat	Joule per kilogram	J/kg
l	Length	Metre	m
λ	Wavelength	Metre	m
m	Mass	Kilogram	kg
M	Moment of a force	Newton metre	$N\,m$
N	Neutron number		
p	Momentum	Kilogram metre per second	$kg\,m/s$
p	Pressure	Pascal (= N/square metre)	Pa
P	Power	Watt (= joule per second)	W
Q	Electric charge	Coulomb (= ampere second)	C
Q	Quantity of heat energy	Joule	J
r	Angle of reflection	Degree	$\circ$
r	Angle of refraction	Degree	$\circ$
R	Resistance	Ohm (= volt per ampere)	Ω
R	Resultant force	Newton	N
s	Distance	Metre	m
t	Time	Second	s
t	Temperature on Celsius scale	Degree celsius	$^\circ C$
T	Temperature on Kelvin scale	kelvin	K
T	Period	Second	s
u	Initial speed or velocity	Metre per second	m/s
v	(Final) speed or velocity	Metre per second	m/s
V	Volume	Metre cubed	m^3
V	Voltage or potential difference	Volt (= joule per coulomb)	V
W	Weight	Newton	N
W	Work and energy	Joule (= newton metre)	J
Z	Proton number or atomic number		

Index